The Inquisitive Pioneer vol III

The book of At-Home Basic-Materials Waves & Astronomy Science Activities solving with a Slide Rule

Bryan Purcell

© 2015

This is for my stellar electromagnetic left-hip, Linda W.

Introduction

This the 3rd book in the series, The Inquisitive Pioneer. Each of them has a cross-section of Activities that all have the same structure and comes from my web site www.cosmicquestthinker.com . Each separately examines a central concept in science and looks through the lens of the Scientific Method. It has a focus of not only observing what happens in a structured and procedural experiment, but each has at its center not only the measurement of data but the analysis of data. This is one of the central goals of science itself and is its operational definition due to the fact that science is really the pursuit of the story of the Universe from the macroscopic to the subatomic to understand the relational structure between all things. Science is the story of what is, and is the fabric woven through the whole of the universe and ourselves. We as scientists, uncover the story of science through qualitative descriptions utilizing science nomenclature and the quantitative depictions from the language of mathematics.

This 3rd book has as its central theme the realm of both Waves and Astronomy. Physics is the primary science where for the most part things are either matter (which has its own set of properties) or waves (which is seen as the means to transmit energy from place to place). Waves are encountered in all aspects of our lives – light is quite obvious and very critical, but so too is sound. Sound is not only what we hear, but also things that we have come to use such as ultrasound (sound frequencies above our sensory level of hearing). Light is only a portion of a large family of waves called the electromagnetic spectrum. In the electromagnetic spectrum we have, measure, and even use waves such as x-rays, ultraviolet rays, infrared rays, microwaves, and radio waves. This spectrum deals with our communication and data transmission. The applications of the electromagnetic spectrum is a list that is ever growing – we use infrared to search for people and animals in the dark, it is used in the study of the heat given off by mechanical system; we use microwaves in cooking and other uses, and so on. It is through these electromagnetic waves that we explore the realm of Astronomy, hence the connection in this book to that particular science. All we know of galaxies, stars, nebulae, planets, and other objects ultimately ties back to waves of one or more types.

In this book, Waves allows the examination of mechanical waves, such as sound and electromagnetic waves, particularly light. Along the way there is exploration of the interaction of waves with matter, such as refraction, and the like. There are labs on optics – how light goes through a material, how it acts when it comes to lenses. There are many experiments on how waves interact with matter, such as heating it up or allowing us through technology to generate electrical current in our pursuit of the use of solar cells.

The philosophy of the book in these Science activities is to encourage an independent thinker and learner. Science is learned by doing. Find the pace you want from the activities. Many activities have a prelude that gives some of the historical and/or mathematical foundations of the exploration. When encountering new ideas, read and reflect on them as well as doing further research. When approaching the activity – read it through and envision it and what should happen first. In setting it up, test it and watch what happens even before measurements. Be willing to redo experiments as needed. With measurements, watch the data, its trends, make mental predictions, estimate the answers sought and the conclusions your experiment is pointing to.

The Science activities in this book are called hands-on, but many books say this. Here the science explorations go from observing to actually doing the science. All activities have as their core the need to have controlled experiments with measurements and the analysis of the data. This is the heart of true science. Observational and qualitative science are the beginning positions for many things, but this book goes the extra step to all things measured, all measurements used for reasoning and reaching conclusions. We become the researcher, the experimenter, and the mathematician.

This is a book for a thinking person wanting to take a personal journey into fundamental science principles through investigation. The primary person it is for is one who is inquisitive, one who not only likes science and math but prefers that science is explored in a hands-on method using basic, even everyday materials. This science-minded person views the world from the scientific method approach. There is the phenomena or question, which sparks one's curiosity to do some research and reading. This then leads to posing a measurable hypothesis. From there the science-minded person gathers materials that can readily be at hand or are easy to obtain for use and be used to reach the goals set forth in the hypothesis. Experiments are ran, measurements taken, and numbers analyzed via mathematics to reach conclusions. All along it is the mind of the student of life that assembles these thoughts, these ideas, these numbers measured and mentally constructs a model of what is.

The many activities that make math a part of everyday life and the science activities use basic mostly everyday items so that a student sees that science is not just in a book or a given classroom but is found everywhere.

The materials in the lists are mostly low in cost – they can mostly be found at home right from the start – such as rulers, measuring tapes, tape, measuring cups, string and others perhaps. If not, then they are readily found in inexpensive resource places such as dollar and dollar & more sort of stores. In some cases there are a few recommended items that can readily be found online with Amazon and the like, and are often low in cost as well, such as a glass prism or a good lab-grade thermometer. There are a couple of activities that use a regular telescope, while one uses a solar telescope as well. However, one of the goals of this book was to keep most items low in cost and readily reusable for another experiment in most cases. When we see basic materials as tools, it encourages our imaginations to consider other ideas and other questions. These basic materials are part of the philosophy of learning explored here – where common items become the templates to build ideas upon in our minds, which in turn not only help understanding but are then easily adapted to new situations as needed. It then allows us to explore other frontiers.

There is a basic reasoning to this book. Besides making science and math easy to do, it brings it to the everyday and brings it home. It allows science to jump off the page of a text book, allows it to leave the classroom, and takes it home and has it spring forth from the mind of the scientifically-minded student. The use of basic materials is to help to create tangible models to think about, recall and learn from in order to grasp key ideas in science. This reasoning also to applies to the use the slide rule as the math tool to prompt the mind to take in the data, the numbers, and use them in the relations between the variables as found in the equations and uncover the relations that are indeed found in nature.

Why then, the Slide Rule?

This opening chapter is my original essay examining the idea of using a slide rule in the classroom of today (see ch. I) – despite all of the available technology. The primary conclusions and goals of the book are quite simple :

1) The Slide Rule though seen by most as an antique is still a valuable tool today in math and science education because it is a tangible tool that strengthens math skills by acting as a bridge between reality and the math that measures it. It enables one to think about and visualize problems; forcing the user to come up with an estimated answer, and the slide rule, itself, yields a reasonable answer that is acceptable in a real-world context.

2) To present a set of basic Science and Math Activities that range in time from middle school to high school using very basic materials, utilizing a recipe of the scientific method as a procedural method to ask a measurable question, use basic materials, engage in a logical procedure, take measurements, and use these measures to do one or more necessary calculations to reach scientifically valid conclusions.

3) Present a brief essay on slide rule history takes a snapshot view of the last 3 centuries (1600s-1900s) where the inclusion of the Slide Rule was a necessary catalyst in the realm of the two major Industrial Revolution periods that affected all nations on Earth and ushered in new sciences, mathematics, innovations and inventions, along with scales of production.

The slide rule, the tangible bridge for the student, can connect all the core classes and is connecting math to the world of science, carries with it history, famous people and places, and promotes active math engagement by the student. This is used in conjunction with the rationale of the opening of the book where science is tangible and can readily be extended from basic materials which promotes models in the mind to understand concepts and which are explored through carefully crafted experiments by being the Inquisitive Pioneer! Resolve, Solve, Evolve! Explore and Enjoy! – Bryan Purcell

Table of Contents

Important Note of Responsibility :

In the case of all of these Activities the following is to be adhered to – All children and students – you need the support, permission, and help from parents, guardians, and/or teachers to do these Activities. Parents and children alike – read ahead through the whole of the Activity so as to anticipate where there may be areas of concern and important levels of awareness. Always employ safe practices when using any items, such as wearing goggles. Be intelligent, mature, and cautious in the use of items – these are tools and you are using them for specific ends in order to learn. Be aware of all concerns – such as the use of hot or cold items, avoid foods that may have allergies associated with them (nuts, et al), sharp objects (knives, scissors, and other edges) – and act appropriately, make smart decisions, and be safe. When done with Activities always put away materials, clean up the area, and then address the calculations portion of the Activity.

Master List of Items

This is a nearly comprehensive list of all the items needed for all of the Activities. Recognize that any one Activity will require only a small handful of these and in most cases you might find alternatives that can act as good substitutions. Most items can be found at Dollar and Dollar and More type stores along with hardware stores. A few items left off the list are either very common or in the other case very exclusive (often an electronic device or optical item) to that particular Activity and will probably only be used there.

Slide Rule	Stopwatch	Paper
Ruler	Marbles	Graph Paper
Meter or Yard Stick(s)	(Slinky) Spring Toy	Pencil / Pen / Markers
Measuring Tape	Desk Lamp	String (kite)
Sewing Measuring Tape	Light Bulbs (various)	Styrofoam Cups
Measuring Cup Set	Penlight Laser	Plastic Cups
Kitchen Mass Scale	Diffraction Grating	Nuts & Bolts (various sizes)
Styrofoam blocks	Ceramic Magnet	Plastic Wrap
Multimeter	Magnet Wire	Aluminum Foil
Index Cards	Scissors	Paper Clips
Thermometers (lab quality)	Solar Cell(s)	Lenses (known focal lengths)
Paper Towel	Stack of Books	Tuning Forks
Flashlight	Paraffin Wax	Plastic sandwich bags
Prism (glass)	Model Electric Motor	Water
Tension Scale	Wires with alligator clips	Cans
Protractor	PVC Piping	Radiometer
Goggles	Tape (regular & duct)	Gallon container
Hole Punch	Elodea	Dowel Rods
Pins	Pop Bottle	Chairs
Use of Indoor Items : Stereo	Outdoor Items : Moon, Sun	Telescope(s) – regular and solar

Ch.I
Why bring the Slide Rule back to the Classroom today?

Using Old Tools to Solve a Modern Problem
The Reintroduction of the Slide Rule to the Clasroom
Bryan Purcell

Originally adapted from my article in
Oughtred Society Journal Volume 19, Number 1, Winter, 2010

In the world of education today there is a greater emphasis on trying to find the best way to create new paths to success in learning. Most of these emphasize not only greater capacity in one's knowledge base but also the employment of technology to achieve these ends. The argument goes that technology is the best inroad since it already exists. Why? Basically the quick conclusion comes from these points : children today were born into this era of technology hence have familiarity with it, and technology is the backbone of tools in use today in the workplace.

Though these seem to be valid arguments, they overlook the most critical part of education : the process of learning is not the end product. For example these tools enable a quick solution to a story problem, but the answer is not the goal of education. It is instead the acquisition of skills to enable problem solving and the employment of these skills. This is the fundamental goal of both science and math education today. It is parallel to saying that results speak for themselves in the scientific process, but it is the assessment of the student's good data acquisition and analysis that is the most important.

Because of this, the original assertion by education today, that the use of technology is a necessity, can be considered potentially invalid and there may be other paths to the goal. One of these paths is the use of the slide rule as a math tool, not only to solve problems but also to act as a visual bridge to force the user to engage her mind, tie the concepts of the ideas, the data, and the formulae to the real world and come up with a reasonable answer to the question at hand.

The first area to examine in this dialogue is why such an interest in math and science education? The answer is obvious – the majority of the available jobs, even in a tense economy, are in the areas of math and science – such as jobs in medical fields, engineering, computer technology, and the like. A National Science Board report of 2008 mentions that the needs in these sectors will be triple that of the rest of the job market. Also, the greater one's level of education, the greater one's chances are to find employment as well as to have lifelong higher career earnings. It could even be argued that one's level of opportunity and flexibility in the market and what level to which one ascends are directly related to one's math and science skills.

Compare for example the math skills needed today to operate point-of-service cash registers that have pictures of hamburgers and fries as compared to the real powerbrokers in a business, not the CEO, but the CFO (Chief Financial Officer). These same parallels exist in the realm of investigatory science and research such as found in entry-level assistants and chief engineers.

America's greatest competitive edge has always been found in its creative efforts in the areas of science and math, which launched many endeavors such as the NASA golden decade of the 1960s to go from leaving the planet, stepping into space, and safely undertaking the greatest mission in humanity's history, the journey to the moon.

A second reason for the importance of math and science education is the fact that there is a great deal of academic competition that is not only national but international in scope. and often the skills of American students have been shown to be lacking in math and science. Numerous reports have shown that American students are regularly taking math remedial courses in college; their scores on national tests in these areas are low, and even basic math skills are lacking (The Final Report of the National Mathematics Advisory Panel in 2008 from the U.S. Dept of Education).

To address the issue, the question then arises, "What approach works best?" The argument posed here will explore both the areas of necessary concern and the application of a solution in the form of the slide rule to act as the best tool to affect one's success. The topics to explore in brief are :

1. Estimation and Basic Math Operations,
2. Simple Formula Manipulation and Understanding,
3. Math Areas of Ratios & Fractions & Proportions & Conversions, and finally
4. the Math Topics of Significant Figures & Scientific Notation understanding and use.

The abilities of Estimation and Basic Math Skills go hand in hand. These need to start early and continue to expand as a student progresses through school. The emphasis should be on the student's acting as the computer and not the machine. Reliance on the calculator shifts responsibility from the person to the machine. The answer magically appears in the window of the machine and it does not include estimation at all. The slide rule, however, necessitates that one must practice and use basic math skills and continually employ estimation in order to answer questions.

How many teachers could tell the anecdotal story of the set of students who ask for a calculator before giving an estimate or an answer and with the machine they ask whether they should multiply or divide the numbers? A slide rule cannot be used unless one begins to master these basic skills and employ a mastery of basic numeracy.

Also realize that the slide rule is a natural extension of practices already in place in most elementary school systems. In order to teach numbers, their relative sizes, and concepts like addition and subtraction, the number line is considered the best visual tool. Unlike all of the other colorful tools which have an entertainment value, the number line yields the answer. In fact two of them places alongside each other help the process of learning addition and subtraction.

The same argument is true for the slide rule. Instead of a linear line, it is a logarithmically spaced line and is useful due to the properties of logarithms for visualizing multiplication and division easily. Notice that it would be the next logical step in education; if the number line works, why cannot the slide rule?

To carry this idea further, the next area of concern is formulae and their manipulation. Most formulae in school are linear (such as area of rectangles, miles per gallon, cost per unit item, density, average speed, force, pressure, and even Ohm's Law) and are readily found on a slide rule. An important note : National Standards have these and many more formulae for which students are accountable today.

To illustrate, take distance as a value on the C scale, set it over the time on the D scale and opposite the D index is the average speed. One scale is one variable and the other scale is the other key variable in a formula. One could easily explore relations quickly and effectively. For example, 'how much time at a given speed will it take to cover some given distance?' and the like. Notice the visual link of distance and time needed for a given speed. One has to read across the scale. Conceptualizing the changing of one variable and its effect on another is very easy having this tool.

The next area for exploration primarily is concerned with proportions and conversions. Here the slide rule wins hands down! One easily can solve proportions faster with a slide rule than one can with a calculator. Also, conversions can be treated as a proportion (as can the aforementioned 3-variable functions). Many studies, too, illustrate the lack of skill in converting decimal values to fractions and vice versa. The slide rule accomplishes this visually and shows all related fractions to a given decimal value instantly as contrasted with the ubiquitous calculator.

Finally, in the area of significant figures and scientific notation, the slide rule is again the master math tool. In the real world, we need typically no more than 2 or 3 digits of value in answer. No one measures a room's length and width, and then calculates the area to the 4th or 5th decimal place when buying carpeting or tiles. Also consider the goal here: to acquire problem solving skills. This being the case, does one really gain by multiplying a number with 5 digits with another one? How the slide rule is of value here is that the typical slide rule is accurate to 2-3 digits despite the size of the number.

This last statement is explained by scientific notation, which is of such a value and is directly related to the slide rule. The slide rule has only the numbers 1 to 10 on a typical C scale, yet in reality it has all the numbers that exist! The user must merely put the number in scientific notation. In math, the multiplication of exponents or division of the exponents is readily handled by addition and subtraction.

There are some important final thoughts on these matters where the slide rule is of great interest. First, recognize that no studies of the slide rule have ever been done, not even as compared to the calculator. In the same line the skeptic might add that it is an antique. In a parallel argument, why then do we use measuring tapes still when there are electronic devices for distance, why not just use a microwave instead of an oven and stove (how many cooking shows use the microwave over traditional oven). And finally, since we primarily use digital clocks, why then continue teaching traditional clock reading?

Ultimately in this idea is the question between the Slide Rule and the Calculator :
Which is best for improving math skills and numeracy? The argument has been presented. It is sound in reasoning and consideration. Finally in this case, to overlook a hypothesis is poor science at best.

Are there other benefits not noted to the slide rule? First, unlike the calculator, the slide rule has an extensive history which can help spark the imagination of presentation and packaging of the ideas about it and its use. It was the most powerful math tool in the history of all handheld devices for 350 years.

Second, the slide rule is directly connected to famous names such as Newton, James Watt, William Oughtred, Joseph Priestly in terms of its construction and use and to those who used it such as Einstein, Hans Bethe, and von Braun as well as including numerous mathematicians, scientists, and engineers.

Third, the slide rule was the first tool outside the human mind used to create most mobile and immobile structures in society such as the Empire State Building, the Golden Gate Bridge, the jet engine, the Panama Canal, and even the Apollo spacecraft.

Third, there are a number of websites that illustrate how to use the slide rule (some in power point format), and no matter the form of slide rule there are no special considerations needed, since the rules for multiplying and dividing do not change despite the style of slide rule. There are even virtual slide rules (see footnotes). Plus there are websites which have slide rule loan programs for a class if a teacher is so interested. Finally one could even download printable scales and have students construct their own slide rule! Imagine making a tool that with the classic 9 scales (C, D, C1, A, B, L, K, S, T) rivals the power of a scientific calculator, is personally hand-crafted, and has such a history. With the basic slide rule, the journey of the mind in acquiring problem solving skills and connecting math and science to the universe, can begin.

Web sites

Information: The Oughtred Society : www.oughtred.org
Virtual Slide Rules: Derek's Virtual Slide Rule Gallery :
www.antiquark.com/sliderule/sim
Information, Virtual Slide Rule, Slide Rule power point
presentation on how to use the slide rule, and printable
scales for making a slide rule : www.sliderulemuseum.com
Slide rule plans
Scientific American magazine reference from May 2006
article on slide rules by Cliff Stoll :
www.scientificamerican.com/media/pdf/Slide_rule.pdf
Luis Fernandes, Dept of Electrical & Computer Engineering,
Ryerson University :
http://www.ee.ryerson.ca/~elf/ancient-comp/
sliderule.pdf
Circular Slide Rule by Dr. Charles Kankelborg, Dept of
Physics, Montana State University :
http://solar.physics.montana.edu/kankel/math/csr.html
Math & Science Activities : www.cosmicquestthinker.com

Data Analysis Math Tool Alternative Consideration

It is agreed upon today that students need to have connections to the ideas they learn and hands-on activities are the first critical step. The next step then is for them to take their measurements and find a way to connect the numbers to concepts. One of the most important goals of science is to **analyze data** to reach mathematical conclusions and find relations between the variables.

The question then becomes : Is there a different way to examine data? An interesting approach would be one where the students are not only acting as the scientists taking measurements, but also as the mathematician analyzing their measurements. **The answer is the tool, the 'stick' with numbers on it.**

What if, students were to use only low-cost basic tools (rulers, meter sticks, string, thermometers, small masses, marbles, a personally constructed incline made of meter sticks, stopwatches, mass scales, etc) for measurable labs. With basic tools the students take measurements themselves and then with the help of the laws of mathematics and through the use of a 'stick' with numbers on it, the students come to discover and find the relations that they can then read about in their texts?

Even in the case of non-measurable labs where the students are merely supplied, straight-forward data, the students can use the very same mathematical 'stick' and find their relations through some graphing and basic computations.

What 'stick' is this? It is the common slide rule!

Why this tool? The **slide rule** is a tangible and visual bridge connecting numbers to the measured real world. It can be seen as an extension of the use of number lines in their early school journey where they were used for adding and subtracting, only here the slide rule is now used for multiplication and division. The slide rule can also act as a motivation for reasoning and mastery of math.

To use a slide rule, one must first estimate answers mentally, know what and why the measured values used are, sequence the mathematical steps of the problem, and understand their place values through scientific notation of both the variables and the answer.

Learning to read the graduations on the slide rule, (along with learning to use a new tool for calculation) is useful in itself. *Hence, the student becomes the measuring scientist and the computing mathematician simultaneously once again, like those long ago who used such tools.* **The most critical present-day problem, then, is to find such a tool.** The references at the end of the article note the International Slide Rule Museum web site, where there is a student-loaner program. For the cost of about $11 per semester, a teacher can be loaned a classroom set of slide rules. There is a power point on how to use a slide rule, along with ideas on its history, and a way to have medals for slide rule competitions as well. Also in the references is a web site, Cosmic Quest Thinker, for suggestions for many science and math activities using slide rules. Each of these has further links to other web sites for virtual slide rules, printable slide rules, publications, even places to assemble one's own classroom set of slide rules and the like.

Data Analysis with a Slide Rule :

In any and all lab situations or even tables of data cases, the students take recorded (or given) data and then merely convert the values into log values of these numbers (read the log value on the L scale from the data value on the D scale on a slide rule). Now they proceed to graph a log-log plot of each of the variables, such as :

- log(displacement) vs. log(time) for constant acceleration cases;
- log(period) vs. log(distance) for pendulums or planets (Kepler's 3rd Law);
- log(Force or Intensity) vs. log(distance) for inverse-square laws (such as gravitational or electrostatic forces or light intensity), et al).

In this new log data set on its graph, draw a best fit line through these points, and then find the slope of the line. The slope taken as the ratio of two simplified whole numbers will show the exponential relation between the variables and the exponents involved.

For example, in the case of constant acceleration (dropped objects or masses on inclines), the ratio of displacement to time will have a ratio of 2 to 1, hence $d \sim t^2$. This means that the graph is of the form $y = x^2$, which is, indeed, a parabola.

Once the variable relation is found, the slide rule can then be used to then check it as well as explore the relation. Continuing the above example, graph now displacement vs. time-squared as well as the log values of each and again for each draw a best fit line and determine slope. The former determines acceleration while the latter should demonstrate a slope of 1.

In the case of inverse-square laws (gravitational, electrostatic), the slope on a slide rule has a ratio of -2 to 1 for Force to Distance. The negative slope is a negative exponent, so it can be seen as $F \sim \frac{1}{d^2}$. This inverse-square law idea applies to light intensity as well.

Even a situation as complex as Kepler's 3rd Law can be examined this way and one finds what Kepler found (using logarithms, no less) that the period-squared is proportional to the distance-cubed for a planet ($P^2 \sim D^3$).

Note that each of these and many more calculations can be done with as simple a tool as a common 9-scale slide rule! This very tool is as powerful as a conventional scientific calculator today.

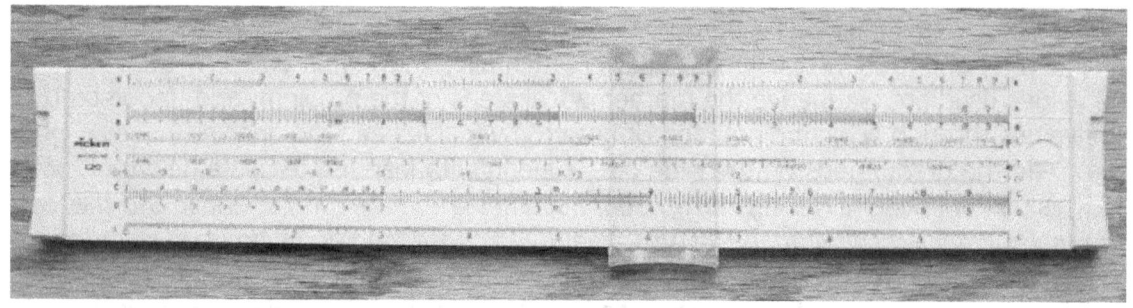

Other mathematical reasons for the slide rule : The slide rule can also be used to illustrate the idea of the *laws of logarithms*, such as the product rule for logs where the log of the product of two values is the sum of the logs of each of the values in question. (log(A*B) = log(A) + log(B)). Students can compare given values and reach a conclusion here. It is a means to visually conceptualize ideas, such as what happens to variables when one of them changes. **The scales themselves become the variable under consideration.** Take, for example students given the density of a pure substance. A student places this value on the C scale opposite the left index of the D scale. Now as they read along the C scale, these values represent mass, the numerator of the equation for density ($\rho = \frac{m}{v}$), while the adjacent D scale is the corresponding volume value for that given amount of mass so as to always end up back at the material's density! Other types of data analysis can be done this way.

Average Speed is similarly done. Distance is the C scale while Time is the D scale. For any determined average speed, as one reads along the C scale, one has driven farther, hence more time (D scale) too. Students can be given data here as well to examine, if preferred.

Because *all similar ratios are set up instantaneously*, this same tool can also be used to easily convert fractions into decimals as well as solve any and all proportions even faster than one can on a calculator. Here tables of information can have blanks to be filled in where students can use the slide rule to find the answers. This can be useful for scaling drawings and maps, calculating changes in recipes, determining cost per unit volume or mass, finding unknown sides of similar polygons, and calculating conversions. **The applications are limited to one's imagination and mathematical skill.**

Notice how this idea extends to a simple activity connecting ideas in math and science in the understanding of the value of π. Students can measure circumferences and diameters of common circular objects and find the ratio on the slide rule. It will show π (3.14) if done correctly and since all similar ratios are set up, for a given diameter (or circumference) they can predict the circumference (or diameter). This application applies to any known ratio. Other scale explorations of the slide rule allow for examination of squares and square-roots (A & B scales), cubes and cube-roots (K scale), as well as trigonometric relations of sine, tangent, and cosine (S & T scales). In combination these scales are all that is needed for virtually all formula in science and math through school. These can all be done as given tables or through measurements depending on the resources and time.

With this approach, using the slide rule, the goal of having the students do the work and discover the outcome is achieved here. The goal of data analysis is achieved. When they do a lab and take the measurements, they now take the data and find the relations using math reasoning when using the slide rule. The students here engage in the art and act of discovery through actually doing the math. The students come to find the various relations either through measured or as given data tables. Along the way, they connect the numbers to real-world phenomena.

Also a startling notion develops – *all values measured can be represented as a number between 1 and 10*, as does the slide rule and this promotes the use of scientific notation. Image their surprise when they realize they are holding infinity in one's hands! The use of the slide rule is just an alternative and a way to inspire a path to mathematical reasoning and understanding. Also consider that the slide rule has a sufficient level of precision with 2 or 3 significant figures, which is all that is needed. The tool helps in reinforcing this idea.

Does this mean the end of the computer or calculator? No. *In fact, the calculator and the computer can act now as a follow-up to check the answers. Instead of being the source of the answers, they are the checking system for the student's work as a follow up.*

What of the use of logarithms and the need to explore them? This can be done in the science or in math class, if the students are at that level for understanding. Otherwise, letting them know that logarithms are a tool to uncover such relations may be sufficient at this time. As noted here, this idea can be extended to any and all other variable relations they encounter in various science classes as well as math classes.

This exploration can be a cool math tool adventure. Students mentally and mathematically examine data themselves to find the answers. They use tools that help them visualize the concepts and make finding answers a personal responsibility and journey. The slide rule promotes math skills acquisition. As an aside, the students can also be introduced to and connected to history through the role of the slide rule. The slide rule was in the hands of numerous scientists, mathematicians, and engineers and used for nearly 350 years (1620-1970). It has a history of being part of the making of the Panama Canal, the Empire State Building, the Golden Gate Bridge, along with development of the steam engine, the discovery of oxygen, and the determination of the density of the Earth. Both Einstein and von Braun used the same 9-scale model themselves – one from the realm of theoretical physics while the other in the practical realm of applied physics to rocket engineering, where he built the Saturn V, the largest human-made device to leave the Earth carrying aloft Apollo astronauts to the Moon, each carrying a Pickett 600 slide rule.

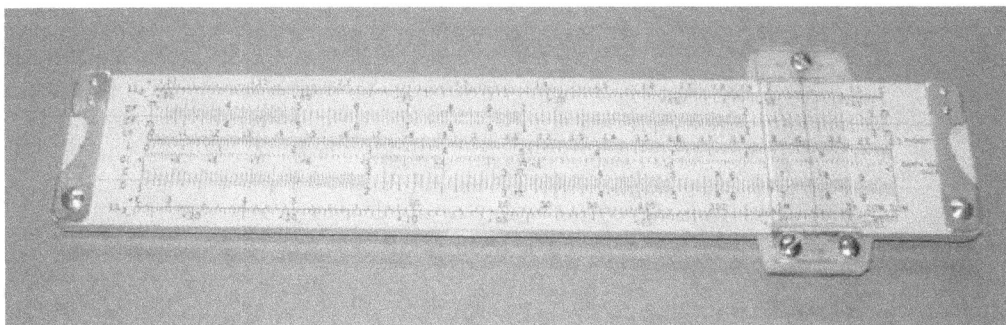

Resources :
Slide Rule Loaner Program, Directions for Slide Rule Use, Make your own slide rule :
http://sliderulemuseum.com
Many Classroom Ideas for Slide Rule use :
www.cosmicquestthinker.com

How does a Slide Rule Work

The Slide Rule is a mathematical tool that enables the user to perform mathematical calculations of a great variety and obtain a reasonable answer. Though it can be used to add and subtract, it is better just to do this oneself. The Slide Rule is best suited for multiplication and division primarily. But with additional Scales the Slide Rule can be used for other things like squaring, taking the square root, cubing, taking the cube root, determining the power or root of a given expression, determining the log of a given number and the inverse of this by finding the number for a known logarithm and in trigonometry can be used to find the values of the sines, cosines, and tangents of given angles along with their inverses where one has the sine of a given angle and needs to find the angle itself. The range of application mostly depends on the user, her math skills, creative but mathematically sound approach to a problem and the number of scales the Slide Rule has to ease the outcome of sought after answer.

The basic parts (and these are noted for the linear model) are the following : The top and bottom strips are called Stators and respectively are referred to as the Top Stator and the Bottom Stator. These pieces are also called Stock in some books. The moving piece between them is the Slide. The Moving Cursor is simple called the Cursor or sometimes called the Indicator or Runner while the cursor line is also known as the Hairline (it was a hair long ago and first suggested by Isaac Newton).

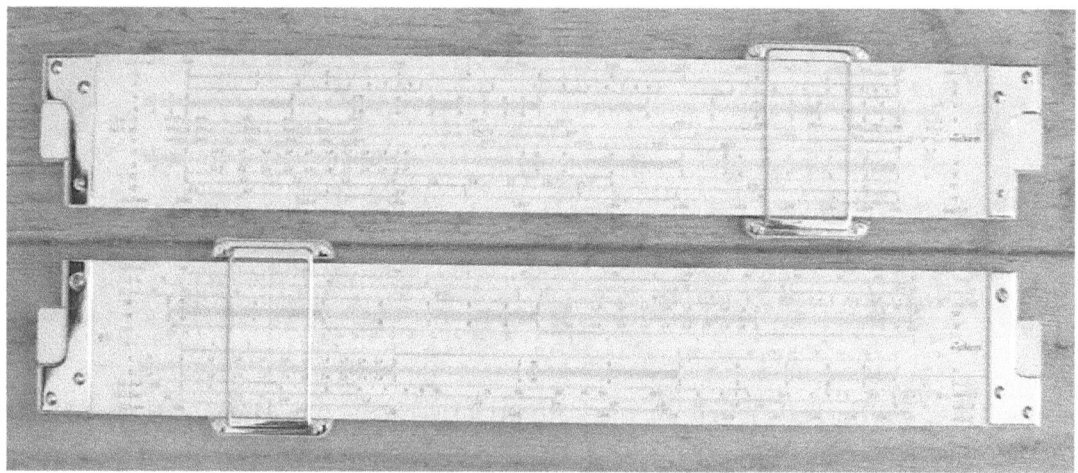

The accuracy of the Slide Rule's answer depends on the Precision of its Scales. Most often the average 10" is effective in its precision of values on its Scales for calculations involving 2 significant digits, but the 3rd sig fig can be estimated as well. The number of gradations on the Scales depends on length, so the longer the Scale the larger the number of gradations, hence a greater level of precision can be found. However, it is not as simple as it might first appear. A 20" slide rule merely has a marginally larger number of gradations hence only a slight increase in estimating the 3rd significant figure. To illustrate : between the 9 and right index 1 of the 10" slide rule there are a total of 10 secondary marks, so values like 9.1, 9.2 are easy enough. Between each of these marks in an ever-narrowing gap (they are spaced logarithmically) on the 10-inch rule there is a mid-point mark, so this would be 9.15, 9.25, et al. It is easy to conclude that one can estimate values to the nearest 0.01, so values like 9.12, 9.37 can be determined where the 3rd digit is estimated while the first two (assuming the slide rule is accurately built and the user is mathematically adept) are reliable. In the case of the 20" rule the same primary and secondary marks exist, but since the scale is longer the tertiary marks now are not 0.5, but instead 0.2. One now has a greater certainty in determining whether the value is indeed 9.13 or 9.14. Notice, however, there is no gain in the number of significant digits though the length of the slide rule has doubled! In general, however, most calculations in the everyday world require only 2 significant digits, so the 10" is sufficient.

The Slide Rule is not a measuring tool, like a ruler, but instead has Scales of Numbers on it that are spaced based on the C & D Scale values which are numbers (such as 1, 1.2, 2, 3.5, et al) placed at a distance from the leftmost number '1' value (aka the Left Index) that corresponds to the logarithm of that given value and multiplied by the size of the scale (in a standard rectilinear rule about the size of a ruler this would be 10" or 25 cm). For example, the number 5 is at a distance from '1' that is log(5)*scale length of the slide rule. This means that all the values are logarithmically spaced from each other on the Scale. What value does having this spacing of values have?

To explore why the numbers on the key base scales (C & D) are logarithmically spaced, first we need to explore logarithms themselves. The reason that it makes the Slide Rule a math tool for multiplication and division comes from the properties of Logarithms, which are :

$$Log(A*B) = Log(A) + Log(B)$$

$$Log(A/B) = Log(A) - Log(B)$$

What this means is this : The log of the product of any two values is merely the sum of the logs of each of the values independently. The log of the division of any two values is the difference of the logs of the respective values as well. What this means is that multiplication can be turned into addition and division can be turned into subtraction. So all we would need then is a table of the log values for any set of values we wish to multiply and we merely add them and find this sum on the table and that would be the product of our values! But the Slide Rule is far easier than this. As noted in the History of the Slide Rule section William Oughtred explored the idea of logarithms and placed the numbers from

1 to 10 at distances from each other that corresponded to the logarithms of the numbers on this line. Knowing the property that the product of any two values is the sum of their logs, all one has to do is combine the distances, hence add them that separate the numbers physically for any given product and the overall total distance will then end up on the number which is the product of the two numbers. Here is a simpler illustration. If I wanted to find the product of 2*3, I merely take the log(2) which equals 0.301 and the log(3) which equals 0.477 and add them together to obtain 0.778 which is the log(6). On the slide rule I merely have to place the starting point, the Index beneath the number in question on another logarithmically-spaced scale – so I place for example the Right Index 1 on the C scale above the 2 on the D scale (go ahead and do this). Now read along the C scale to 3 and look back to the D scale and what do you find? 6, of course. You have traveled the distance of 2 on the D scale and the distance of 3 on the C scale – adding these values together is the same as the product of the values since they are logarithmically spaced so we are to the answer of 2*3 or 6. Read more thoroughly the section on Using the Slide Rule to gain greater insight into using the slide rule for basic and more advanced skills as you learn more and do more with it. Notice we do not need to know logarithms themselves, how to derive them, find tables of them, and with a slide rule in hand, we do not have to construct our own math tool to do this.

Reading the Scale on a Slide Rule :

- All slide rule forms (linear, circular, et al) will have the same basic method in reading the scales
- A Scale is the logarithmic-based spaced numbers or a related line to the base scales (C & D)
- Each mark on a Slide Rule can represent any value as needed through Scientific Notation. So only the values 1 to 10 are needed.
- Watch when doing division or multiplication with scientific notation – see those rules if needed.
- Essentially when values are in scientific notation and are multiplied, then the exponents are added. When values are in scientific notation and they are divided, the divisor (the denominator) is subtracted from the quotient (the numerator).
- All scales are related to the base scales (the C and D scales)
- A base scale has the values from 1 to 10, logarithmically spaced.
- However, each mark has a particular reading based on its place on the particular length slide rule.
- The Numbers on the Scale are the Primary Marks. The next major sets of marks are located between them are called Secondary Marks, and when marked, the marks between these are called Tertiary marks.
- When there are 10 marks between any marks, they are tenths (0.1, 1/10)
- When there are 5 spaces between major marks these are two hundredths (0.02, 2/100)

- When a cursor is between any set of corresponding marks, this value becomes the estimated digit and can be done to the level of 0.1 of the major marks involved at best. If very small distances, then it is best to assume only 0.5 of the value between marks.
- The Primary Mark on all scales is the Index which is the number 1. There is a Left one and a Right one on linear slide rules.
- Linear scales (C & D) have 2 Indexes while circular ones have 1.
- The Indexes on a linear slide rule are referred to as the Right Index and Left Index.
- The majority of slide rules have capacity for 3 significant figures in their calculations. (the range is 2 to 4)
- Best General Rule : Use Scientific Notation in computations
- A, B, C, D, C1, CF, DF, C1F, K, and R scales are all logarithmically spaced values. Some are one time (C & D), others (A & B) are double, some are triple (K), some are reverse (C1 & C1F), some are common square roots (R), some are folded (i.e start at a different point) (CF & DF – starting at π)
- Trigonometric Scales include : S, T, ST
- Log Scales include L, LLN(+ & -) Scales – L is the log value (usually base 10) of a value while the LLN scales are representative of natural log based powers of a C or D scale value to a given exponent (see LLN scales for more information).
- Note, unless noted basic math is done with C & D scales
- Scales written in black are from left to right, while in red are right to left
- All basic scales begin with 1 (except for folded, trig or Log scales).
- Scales are aligned to read across them.
- When a Slide Rule is a duplex (two sided), the cursor can be read for both sides as needed.
- The last digit is estimated in reading a slide rule.
- It is necessary to keep track of the decimal place in calculations.
- Some of the most critical skills in using a Slide Rule are these :
    1. Always mentally keep track of an Estimated Answer (some sort of range for the answer)
    2. Keep track of the Decimal place in using the slide rule, since it has infinity contained in 1 to 10
    3. Mentally visualize the formula under consideration and project this onto the slide rule so as to keep track of which scales to read

Multiplication with a Slide Rule :

- Quick summary of rule (explained step by step below) : Set one index of the C scale to the multiplicand on the D scale. Next, set the cursor of the runner to the multiplier on the C scale. Finally, read the Answer on the D scale under the cursor hairline.
- [Note that the answer can be found on either the C or D scale and it depends where one starts and which index is used – it is up to you, but is determined by whether you go off scale – If the calculation is not present, then use the other Index].

- To multiply with a Slide Rule set the Index of one scale (C) over the first number to be multiplied on the opposite scale (D). [Think of this as X from Q = X*Y]
- Slide the cursor to the second value as it appears on the index scale (C). [This is Y from the equation]
- Read the answer under the cursor's hairline on the opposite scale (D). [This is Q]
- Numerical Example :
- Place the Left Index of the C Scale over '2' on the D scale.
- Now read along the C scale to the value of '3'.
- Opposite 3 on C is 6 on the D scale, which is the answer of 2 x 3. This is because you have moved the scales relative to each other both the distance of log(2) and log(3) when added they are the log(6) distance, hence 6 is the answer.
- What is being done is the addition of the distances of the logs of the values in question
- (Log (X*Y) = Log (X) + Log (Y))
- Though illustrated with C & D scales, note it can be done in reverse (that is alternate C for D and vice versa) and this action can also be done on any paired scale-slide combination, such as A & B or if preferred CF & DF.
- With both multiplication and division (plus other functions) keep track of the decimal point. Using Scientific Notation is the best choice.
- Scientific Notation Method :
-
- In Scientific Notation, when multiplying the exponents are added.
- Special Rule : If the slide projects to the Left when performing multiplication, Add One to the Exponent Value for the correct answer.
- (Note : This is for each calculation, so if done twice, then add two for example)
- (Note : Also very important these rules apply to only the use of these scales in use, C & D, for example and does not apply to reading across to other scales, such as CF & DF for example)
- In Scientific Notation, when dividing the exponents are subtracted. (divisor exponent – dividend exponent).
- One must note, however, whether or not the coefficients when multiplied exceed 10, etc.
- Special Rule :
- Also if the slide in division projects to the right, then the answer from scientific notation exponent needs the subtraction of one from the answer to make it correct. – Note this is for C scale as the Numerator and D scale as Denominator. If the scales used are done in reverse of this, then so too is the rule, hence if it projects to the left in that case then subtract one. (The idea here is the same as it was for multiplication – the number of times the slide extends to the right, for each time – subtract one from the exponent total).

-
- Key to calculation is know your decimal placement from products and ratios and using scientific notation it is always one to the right or left of number on the slide rule!

Division with a Slide Rule :

- Quick summary of rule (explained step by step below) : Set the cursor hairline of the runner to the dividend on the D scale. Then slide the divisor on the C scale under the cursor hairline. Finally, read the Answer on the D scale under one index of the C scale.
- [It is important to recognize that for both multiplication and division the use of the scales can be reversed. Here in step one, C is the numerator while D is the denominator. These roles can be reversed].
- To divide with a Slide Rule set the divisor value on one scale (C) over the dividend value on the opposite scale (D).
- Read the answer above the active (D) scale index (1) on the opposite scale (C).
- If the answer falls outside the range of the scale (such as when multiplying or dividing numbers from opposite sides of the scale) then the other index needs to be used in a linear slide rule.
- (Note : This does not occur with a circular slide rule).
- Numerical Example :
- Place '6' on the C scale over '2' on the D scale.
- First note that we are creating a ratio of 6 over 2.
- Next realize that we need to use the Index. In the case of division, always use the Index on the Scale that is the Denominator – here that is the D scale for the example, but it can be the other way if you wish.
- Now notice that the Right Index of D is not opposite any value, so we need to examine the Left Index of D, which is opposite the value of '3' on the C scale, the answer to 6/2.
- What if we wanted 2 divided by 6?
- Here the denominator is the C scale, so again find the useful Index, which is the Right one in this case.
- Since 2 < 6, we expect our answer to be less than 1.
- The right Index of C is opposite 0.333 on the D scale.
- Notice that 3rd digit – it comes from the fact that each division between 3.3 and 3.4 is 0.02 and the cursor is in the mid point between 3.32 and 3.34. Finally following the rules of scientific notation and decimal placement, since the slide went to the left and each of the values has an exponent of 0, so 0-0 is 0 and then we subtract one from that and have - 1 for our exponent, so our answer of 0.333.
- What is being done is the subtraction of distances of the logs of the two values in question
- (Log (X/Y) = Log (X) – Log (Y))
- The best way to consider this operation is think of it as a proportion : the Answer is over 1 while along the scales the Divisor is over the Dividend.

-
  - The Proportion concept is useful for calculations involving conversions and manipulation of 3-variable functions.
  - With all slide rules where one scale can move while the other is immobile all answers are present in both multiplication and division as the scales are read. This makes it a parallel calculator – all relations are instantly set up and visual.
  - Though the Scientific Notation method works best for values, be wary and if the slide projects to the right and you are reading the Right Index, then in using the Scientific Notation system for the values, subtract one from the answer in all cases. If it projects to the right, then it is alright for division, yet remember that for Multiplication you add one to the total exponent value!
  -
  - Alternative Multiplication and Division Method :
  - Characteristic Method (which is akin to the Scientific Notation method, only here the characteristic is the exponent) :
  - For any given set of values, write the characteristic (the portion in front of the decimal when written as the log of the value in question (recall the notation : characteristic.mantissa – Note only the characteristic is needed, we are not looking up the mantissa, which, by the way, can be found on the L scale of your slide rule if needed) –
  - Be sure to be wary of positive and negative values in this case!
  - Sum up these characteristics.
  - Now perform the Multiplication or Division with the Slide Rule as you normally would. That is to say, each number is merely a value from 1 to 10 only. Note that the answer may end up on the range of values before 1 to 10 (i.e. be 1/10 as large) or on the range after 1 to 10 and be in the range of 10 to 100.
  - For multiplication, each time the slide , with the C Scale, extends past the left index of the D scale, add one (+1) to the Sum of Characteristics.
  - For Division, subtract one (-1) from the Sum of Characteristics.
  - The revised total is now the characteristic (i.e. the power of 10) for the answer to use with the value showing on your slide rule.
  - Return to our 2 x 3 and 6/2 examples for a moment. Each has a Characteristic of 0. In the case of multiplication, it went to the right, so the sum is again 0, while in division the sum is 0, but we subtract one yielding an exponent of -1 from these rules here.

Combined Operations with a Slide Rule :

- Combined operations are multiple operations in one problem
- In this list there are many examples and notated use of the Slide Rule when it comes to fractions, ratios, proportions, and applications of these ideas.

-
- Let's say we have the situation for continued products where
  Q = A x B x C x ---
- Here set the hairline of the cursor indicator at A on the D scale.
- Next move index of C scale under the hairline.
- Next move hairline over B on the C scale.
- Now move the index of C scale under the hairline.
- Next move the hairline over C on the C scale
- Now move the index of C scale under the hairline.
- Continue moving hairline and the index alternately until all the numbers have been set in this case and come to the answer.
- Second scenario :
- If the problem reads (N x M) / R then
- Place N (C scale) over R (D scale)
- Slide the cursor along the D scale to M (on D)
- Find the Answer on the C Scale
- Repeat this process as needed for more than this set up
- Be sure to keep track of Decimal Point as noted above in the Multiplication and Division Rules.
-
- In essence, combination problems can be seen as proportions which are extensions of ratios, which slide rules are good at.
- For example, any two values over each other is a Ratio or Fraction and once set all similar ratios are automatically established instantaneously. As well opposite the Index of the Divisor is the decimal equivalent of the ratio as well!
- Convert the decimal to a fraction -
- <u>What of converting from a Decimal to a Fraction?</u>
- Place the decimal value on the C Scale over the Right Index of the D Scale and then search along the C and D Scales for a ratio of numbers to represent them.
- Keep in mind this decimal is the percentage value, but you need to multiply in your head by 100 to see it as a percentage.
- For example, 3 on the C Scale over 8 on the D Scale has the Right Index of D under 375 on the C Scale.
- Since 3 < 8, the value is less than one and should be read as 0.375
- What of the Scientific Notation Rules. Each initial value has the exponent 0 and in division, 0 − 0 is 0. But the slide is projecting to the right so we subtract 1 from the result and have -1 for an answer. This means to move the decimal one to the left, and the answer is read 0.375
- As a percentage, multiply by 100 and the answer is 37.5%
- <u>One of the largest uses of fractions is for Sales!</u>
- Place the Price of the Item on the C Scale over the Right Index of the D Scale.
- Read backwards along the D Scale to 9, 8, etc. Each of these is read as N x 10% (such as 90%, 80% and so on). Above it is the Price at that percentage!

-
- Of course, keep in mind, this is not the percentage off, but what you are paying. To see how much is saved, just use the Left Index of D Scale under the Price instead and read to the right instead of left.
- For example, 25% is found at 2.5 and the value above it is 25% of the price. (Of course what you pay is found at 7.5 or 75% instead).
- Sales tax and total cost can be found in a similar manner :
- Place the Cost of the item over the left index of the adjacent scale 9 say C over D as we have been doing)
- Read along the D scale to the sales tax expressed as a decimal value (this should not be too far, 4% is 0.04, 6% is 0.06, et al)
- The value above on the C Scale is the total cost including sales tax!
-
- Other Fractions or Ratios in Everyday Life :
-
- Mpg = miles per gallon
- Take the ratio of miles driven over the number of gallons used sometime to determine just what gas mileage you are getting!
- Mph = miles per hour
- What was your average rate of travel for a trip?
- Take the ratio of the Miles driven to the amount of time (in hours) to find the average speed of the trip!
- What if you are some fraction of the distance there and have determined the average speed, then
- The question becomes how much longer 'to grandma's house?'
- Take the remaining distance and divide by the average speed to estimate the number of hours to complete the journey.
- Further, from the mpg calculation you could take the remaining distance divided by the mpg and find the number of gallons needed for the trip in order to decide whether to fill up again or not!

- The fraction as a Slope in Algebra
- A ratio might not be just two numbers, but instead represents the ratio of two differences of numbers :

- $m = \dfrac{\Delta Y}{\Delta X}$

- This is the slope of a linear line, one of the largest topics in algebra.
- The slope is the rate of Rise Over Run. The larger the value, the steeper the line.
- The sign of the slope determines whether the line is moving up or down when examining it from left to right.
- You can consider slope as a 3-variable formula (see below).

- The Fraction as a Rate :
- In many practical applications of the fraction it is seen as a **Rate**. The numerator is one value (distance, gallons, cubic feet of gas, amount of

- growth, temperature change, etc) while the denominator is some other value (typically time).
- The rate could be determined or it may be known in a given problem. If to be determined, the amounts of the other two variables are known
- If the rate is known, then clearly we are missing either the amount in question flowing or the amount of time needed to do this.
- A very good example of a <u>Rate is in Cost per Unit Ounce (Volume)</u>, etc. This is a valuable tool for the slide rule in comparing items when shopping for their comparative costs to find the better deal.
- The flow rate is how water and natural gas consumption is measured and billed. For water and natural gas are Ccf (100s of cubic feet).
- In order to estimate the Cost, one has to merely read the meter at the beginning and the ending times, subtract to arrive at an amount used and multiply this by the cost per unit. This is true for all of the meters : Water, Electrical, and Gas.
- <u>Rates and Ratios are very common in Science :</u>
- In Physics there are many ratios, such as Density (amount of substance per unit volume), Pressure (force per unit area), Power (Joules per second), etc.
- Also in Physics, rates can be speed (rate of change of distance with respect to time) or acceleration (rate of change of velocity with respect to time) or from Newton's 2nd Law acceleration is the ratio of Net Force to the mass of the object undergoing the net force.
- Chemistry has many ratios such as Molarity (the number of moles of solute to liters of solution), Molality (the number of moles of solute to kilograms of solvent), Percent by Volume [Mass] (the volume [mass] of Solute to the volume [mass] of solution), the Law of Definite Proportions where the Percent by Mass (is Mass of the Element divided by Mass of the Compound), as well as the Law of Multiple Proportions (these could be looked at in the Proportion section obviously), and so on.
- In Chemistry rates are seen in things such as rate of reaction (is the negative of the rate of change of reactant to change of time) plus many more.

- <u>Conversions :</u>

- In Conversions there are two basic methods that can be used with the Slide Rule :
- First, treat the two items as a Fraction that is to be multiplied by the Number in question.
- Typically the ratio of the items is taken where the units one is 'in' are in the denominator, while the units one wants to convert into are in the numerator.

- $$\frac{\textbf{Units to Convert Into}}{\textbf{Units to Convert From}} \textbf{ or } \frac{X}{Y}$$

- This results in this situation :

-
- Beginning Value in Initial Units* $\frac{\text{Desired Units}}{\text{Initial Units}}$ = Final Value

- **N * $\frac{X}{Y}$ = M**

- The best way to handle this on a slide rule is to :
- Place N on the C Scale over Y on the D Scale on the slide rule
- Read along the D Scale to X, the Desired units and read the answer, M on the C Scale.
- Looking at the prior discussion, it is easy to see that the formula presented can be read as :

- **$\frac{M}{N} = \frac{X}{Y}$**

- All one has to do is take the ratio of Convert Into Units on the C Scale over the Convert From Units on the D Scale
- Now read along the D Scale to the Beginning Value (N) and find the answer above it on the C Scale (M)
- Use the following Table and look elsewhere for everyday conversions. It is best to keep a small list and memory of these as needed :

  - **1 inch = 2.54 cm**
  - **12 inches = 1 foot**
  - **3 ft = 1 yard**
  - **1 mi = 5,280 ft**
  - **1 minute = 60 seconds**
  - **365 days = 1 year (rounded)**
  - **1 solar day = 24 hours**
  - **1 cup = 8 fluid ounces**
  - **1 gallon = 4 quarts**
  - **1 pound = 16 ounces**
  - **1 kilogram = 2.2 pounds**
  - **1 ounce = 28.3 grams**
  - **1 liter = 1.06 quarts (rounded)**

- Plus look up whatever you may need !

- For each of the above and any others your find a need for, simply place one value over the other as described above it for use.
- <u>Conversions are needed in many applications :</u>
- One basic unit to another (inches into feet),
- Changing one unit type into another (English to Metric),
- Currency Conversions, and others.
- <u>Computing Costs</u>

-
- The reason for conversion depends on the needed outcome. For example feet into yards is commonly used when computing the area of a room in square-yards for carpeting by dividing by 9.
- In the case of painting, the Slide Rule readily calculates the area of a wall with Length times Width, but what of the number of gallons of paint needed?
- Take the total area to be covered and use the Slide Rule to divide by 300 if the walls are unpainted or rough – if smooth and already painted divide by 350 – to determine the number of gallons of paint needed!

- In Math, angles are often converted between
  Radians and Degrees :

- $180° = \pi$ radians

- In the Sciences, there are numerous conversions not only of the aforementioned units, but also of mixed units :

- Such as km/hr to m/s is very common.
- For that calculation the ratio for it is 3.6/1. Check it yourself!
- In Chemistry there are conversion commonly found in the amount of substances and how it is expressed :
- For example take the number of grams of a substance and divide by its gram molecular weight to determine the number of moles present.
- In Science and Math there are many other conversions depending on the situation at hand.
- In Math in the realm of trigonometry since the C and D Scales are the values for the Sines and Tangents as read from the S and T Scales (keeping in mind where the decimal falls), this makes it easy to multiply or divide by the sine or tangent of a value when and where needed.
- Slide the sine value for a given angle read on the C Scale from the S Scale over a given value on the D Scale.
- Read in one way the sine is in the numerator and in the opposite direction it is in the denominator.
- What if you want to find the Sine or Tangent value and are given the sides of a triangle?
- Obviously this is again a ratio : For example :

- $$\sin\Theta = \frac{\text{Length of Side Opposite}}{\text{Length of Hypotenuse}}$$

- Also other angular measures are available and can be used for determining distance or size, such as in the Radian measures :

- $$\textbf{Radian measure} = \frac{\text{Arc Length of Circle Portion}}{\text{Length of Radius}}$$

- If our protractor measures 1/10th of a radian, then the apparent size of the object in question is 1/10th the measure of the distance from us!

- Another interesting one in Math is the conversion from one base to another in terms of logarithms.
- The standard slide rule has base 10 logs, but let's say you want a number in another base, say 2 or the natural log base e (2.71828*)?
- For any positive numbers, N, A, and B with A ≠ 1 and B ≠ 1,

- $$\log_A N = \frac{\log_B N}{\log_B A}$$

- From the question the natural log of any value is the ratio of the log base 10 of the number divided by the log base 10 of the natural log value.

- To illustrate, here is a similar example question :
- What is the log base 2 of 6, for example. $\text{Log}_2 (6)$
- Look up reading from the d Scale to the L Scale both the log of 2 and the log of 6. (0.301 and .778 respectively)
- Now divide on Scale C and Scale D 778 by 301 - We find it to be 258
- Where does the decimal go?
- Since 6 is much greater than 2, it will have a characteristic (here 2 and the rest is the mantissa 0.58) so the answer is 2.58
- So $2^{2.58}$ = 6 (try it and see, rounded off of course)
- The idea of the ratio, fraction goes on to any and all applications.

- **<u>The Proportion :</u>**

- As noted in the prelude, the proportion is two ratios set equal to each other.

- In the conversions section above, it is easy to see its value in use.
- The basic proportion can be expressed as :

- $$\frac{M}{N} = \frac{X}{Y}$$

- All one has to do is take the ratio of known values - M on the C Scale over N on the D Scale

- Now read along the D Scale to the other known value (Y) and find above it on the C Scale the answer (X)
- It is easy to see that all one needs to know is any 3 of the variables and the 4th is the one to find.
- By using the scaled as fractions it is easy to place one value on one scale and the other on the opposing scale.
- Try this for yourself to find that solving a proportion on a Slide Rule is indeed much faster than one can solve one on a Calculator!

-
- Also here there are no complex rules, like the calculator, such as cross-products –
- In the case of the Slide Rule the natural form is maintained which is the equivalence of two ratios.
- This is an invaluable tool in Algebra for proportions as well as in geometry for any and all similar figures to determine unknown sides!
- Proportions can be used to determine height or distances - In this case we are using similar triangles.
- For example, hold up a ruler at arm's length to measure the apparent height of a distant object, say a picture on the wall.
- If you read the apparent height, measure your arm's length, (this is the first ratio)
- now measure the distance to the wall,
- Set the first ratio of your measures to the ratio of the unknown height on the wall to the distance to the wall.
- Then the height of the picture can readily be determined.
- Though simple, this technique is used in surveying regularly and is used in the wilderness for distances across rivers and gorges before the advent of electronic equipment.
- The easiest way to envision it is when you are outside and you as well as a tree or a flag pole casts a shadow on a sunny day.
- The ratio of your shadow length to your height is equal to the shadow length of the flag pole to its actual height.

- $$\frac{\text{Length of your shadow}}{\text{Your Actual Height}} = \frac{\text{Length of flagpole shadow}}{\text{Actual Height of flagpole}}$$

- A more thought out activity involves determining the size of the Moon :

- Use a meter stick and place a small card vertically with a hole punched in it at a distance
- Look along the stick through the hole so that when viewing the full Moon it fully fills the diameter of the hole (i.e. move the card until you have proper alignment)
- The ratio of the diameter of the hole to the distance that the hole is from your eye equals the actual diameter of the Moon to the Moon's distance from you. (Here, we assume we know the distance to the Moon). Hence the diameter of the Moon can be determined!

- Proportions can also be used in changing the scale of a recipe :

- $$\frac{\text{Recipe Requirement for Material}}{\text{Recipe Number of Servings}} = \frac{\text{The amount of Material Needed}}{\text{Number of Servings}}$$

- Here the ratio of how much is needed to the number of servings is set equal to the amount of unknown material and the number of desired servings. The amount needed is readily found.

- **In Math there are many applications :**

- The very nature of the proportion stems directly from geometry and the relations found in similar figures.
- These can be the aforementioned triangles but also includes any similar polygons, such as squares, rectangles, and the like where one is known, the other is partially known and there is a missing side.
- Still other examples are considered :
- *What if one wants to find the circumference of a given diameter of a circle (or vice versa)?*
- Simply put π on the C Scale of the slide rule over the Index on the D Scale. – Note with CF & DF scales this is already done since these scales are set at π as their beginning point over the C & D Indexes! See CF & DF scale use
- Now read along the known scale. The C Scale is the Circumference, while the D Scale is the diameter.
- *What about the diameter if the area of the circle is known?*
- Take the Area on the A Scale over p on the B Scale.
- The diameter-squared is found on the A Scale opposite the B Index.
- The diameter is then found below reading from the A Scale to the D Scale to read the square root of the value on A.

- Other interesting things can be done with gauge marks (special marks on slide rules for conversions, multiplications, et al like π) as well as personally derived values :

- On some Slide Rules there is a mark, c, on the C & D Scales (1.273) which is $\frac{4}{\pi}$ and comes from $A = \pi * \frac{d^2}{4}$
- Place this C mark or value of the C Scale over the index of the D Scale
- Slide the cursor over the size of the diameter of a considered circle on the D Scale.
- To find the Area of a circle read the answer on the B Scale!
- What if you want to do this by knowing the radius (of course we could simply multiply by 2 to use the former method, but give your mind a chance to explore the Slide Rule !)
- Look up the square root of π by first finding it on the B Scale and noting its value on the C Scale.
- Slide the square root of π value on the C Scale to the Index of the D Scale
- Now read along the D Scale to any desired radius value for a circle.

- The answer in this case is read on the B Scale for the Area of the Circle in question!

-
- What about the Volume of a Sphere?
- Now place the value 1.61(2) on the D Scale over the index of the C Scale.
- Read along the D Scale, for the radius of a given sphere you are considering.
- The Volume of this sphere is found on the K Scale!
- This comes from $(\frac{4*n}{3})^{1/3}$ from the formula $V = \frac{4*n*r^3}{3}$
- There are many other problems in Math and Algebra in the area of problem solving:
- For example, if a given material costs so much per ounce, pound, ton, how much will another desired amount cost ?

- $$\frac{\text{Cost}}{\text{Unit Amount}} = \frac{\text{How much does it Cost?}}{\text{Amount Wanted}}$$

- There are numerous ratios of values that can be found or derived from many other references that can be used in proportions to find the answer to many a question one might encounter in a math and or science text:

- $$\frac{\text{Diameter of Circle}}{\text{Side of Inscribed Square}} = \frac{99}{70} = \frac{\text{Diagonal of Square}}{\text{Side of Square}} = \sqrt{2}$$

- Here is a general value for the pressure one feels with depth :

- $$\frac{\text{Pounds per Square Inch}}{\text{Feet of Water}} = \frac{26}{60}$$

- Even more complex proportions can be solved :

- A classic algebra question might read :
- *If it takes 4 people 7 days to accomplish a job, how much time is needed (assuming the same work rate) for 6 people to do this task?*

- $$\frac{\frac{\text{Initial Workers}}{1 \text{ Task}}}{\text{Amount of Time Initally}} = \frac{\frac{\text{Workers in case 2}}{1 \text{ Task}}}{\text{Amount of Time needed}}$$

- $$\frac{Wi}{\frac{1}{Ti}} = \frac{Wf}{\frac{1}{T2}}$$

- $$\frac{4}{\frac{1}{7}} = \frac{6}{\frac{1}{X}}$$

- You could go through and first simplify it and then take the ratio of the numbers on one side once the variable is isolated, but the slide rule allows for this to be solved as is!?

- Take 4 on the D Scale and slide 7 on the C1 Scale over it. Start with the cursor here.
- Read along the C1 Scale to 6 and look below on the D Scale to find the answer 4.66 days.

- **In Physics, for example, say you have a balance beam.**

- *If on one side of the balance you have 26 g a distance of 32 cm from the center,*
- *how much mass must be placed on the opposite side at a distance of 20 cm from the center in the opposite direction so that it balances?*

- cw is clockwise, ccw is counter-clockwise

- $$\frac{\text{Mass cw}}{\text{Mass ccw}} = \frac{\text{Distance ccw}}{\text{Distance cw}}$$

- $$\frac{26 \text{ g}}{X \text{ g}} = \frac{20 \text{ cm}}{32 \text{ cm}}$$

- X = 41.6 g

- This idea can be applied to problems in chemistry too.

- Take for example, conservation of mass in a problem where one has to determine the mass of a reactant product in a total mass size where one is only given a small sample for testing.

- $$\frac{\text{Mass of reactant in sample}}{\text{Mass of Sample}} = \frac{\text{Mass of reactant in total mass}}{\text{Total Mass}}$$

- This list applies to any and all sciences and is limited only by the imagination in the questions being asked.

Using the CF & DF Scales (πC & πD) :

- The CF and DF Scales are called Folded Scales since they do not start at the Index, but instead are at a chosen point, namely here being π.
- Why π?
- Simple – there are many calculations that involve π that extend from Circles, such as the circumference of a circle :

-
- Take any value on the C or D scale and now look to the cursor value on the corresponding CF or DF scale. It is p times greater. This is the formula for the Circumference of a Circle : C = π*d, where d, the diameter of the circle is being read from the C or D scale and its circumference is then found on the CF or DF scale.
- For example – put the cursor of your slide rule on 3 on the D scale and find the DF scale. We are assuming we have a circle with a diameter of 3 units, so the question is : what is that circle's circumference? On the DF scale we read the answer of 9.42 units
- Note that the reverse is true as well. If we know the circumference of a circle, we can read it on the CF or DF scale and find its diameter quickly on the C or D scale.
- Another important role that a folded scale has is this : Since it starts at another point on the line, it is easier to do multiplication and division without having to change up Indexes as often.
- For example let's try 3 x 6
- We might first be inclined to slide the Left C Index over 3 on the D scale before realizing we should have first chosen the right C Index.
- But no bother –
- Read the value of 6 on the CF scale and find opposite it on the DF scale our intended answer, 18!
- What of the decimal point rules here, however?
- Going back to our original rules,
- First estimate the answer : Anything past 3.x will result in a value greater than 10, since by 3x4 we are at 12 already.
- Next, if we were to only use the C and D scales, then we would have had to use the Right Index instead of the Left one, hence we would have added one to our exponent total, or 1 in this case (since the values started with exponents of 0).
- Another way to think of this problem is this : Since we have gone past the end of the scale we are on and moved on to the next one in sequence, it is 10x larger, so instead of being 1 to 10, it is now 10 to 100, so the value of 1.8 is really 18 in this case.
- Note : You can do multiplication and division with the CF and DF scales and the same rules apply for them as they did for the C & D scales presented previously. Realize that the index is still 1.

Using the C1 Scale ($\frac{1}{x}$) :

- This is the Inverse Scale. – See an example of use in combination problems – this is a common use of the C1 scale.
- It is basically C scale in reverse.
- It is written right to left (regular are left to right)

- The inverse of any number can be found by aligning a cursor over a given value of C and on C1 is the inverse. (Be sure to keep track of the decimal!)
- For example if reading '5' on the C scale, on the C1 it must be 0.2
- To multiply with the C1 Scale set one of the numbers on C1 over the other number on D.
- Read the Answer on the D scale under the Index on C1 scale.
- Any problem with a fractional component can be examined more easily with C1. – It can be a fraction or be the denominator of a given expression.
- Especially useful in combination problems. – see prior for example in the combination section under proportions.
- Use the rules for multiplication and division and read the proper index for a solution
- One of the other important and common uses of the C1 Scale is to be the scale when reading for a value from the tangent scale angles above 45° (45° to about 80°)

Using the A & B Scales (Squares X^2 & Square Roots $\sqrt{X}$) :

- The A & B Scales are double scales and are the squares of C & D Scales –
- That is to say it is double logarithmic (1 to 10 and 10 to 100)
- Conversely C & D are the square root values of A & B
-
- Rules for Square Roots of Numbers :
- Note Left side is Left Index to Middle Index, and Right side is Middle Index to the Right Index
-
- Odd Number Digit Rule :
- For Whole Numbers > 1
- If the number of digits left of the decimal point in the value being considered for square rooting is odd, Read the value from Left-Hand side of A
- For Numbers 0 < x < 1
- If the number of zeroes to the right of the decimal is odd then Read the value from the Left-Hand side of A
-
- Even Number Digit Rule :
- For Whole Numbers > 1
- If the number of digits left of the decimal point in the value being considered for square rooting is even, Read the value from Right-Hand side of A
- For Numbers 0 < x < 1
- If the number of zeroes to the right of the decimal is even then Read the value from the Right-Hand side of A – Note this includes no zero at all (just .X)

- To summarize the rule (and make reading a R scale easier) :
- The Left hand side of the A & B scales is for Odd Number of Digits or the Odd Number of Zeroes in a Number (this corresponds to the R1 scale)
- The Right hand side of the A & B scales is for Even Number of Digits or the Even Number of Zeroes in a Number (this corresponds to the R2 scale)

For 0 < X < 1									
No. of zeroes in between X and the decimal	0	1	2	3	4	5	6	7	Continue the pattern of even then odd values
Which Side (L or R)?	R	L	R	L	R	L	R	L	Continue pattern R,L,etc
$\sqrt{}$ Answer and the no. of zeroes between ans. And decimal point	0	0	1	1	2	2	3	3	Continue Pattern 4,4,5,5,etc

Number to have a square root taken	Answer on slide rule
0.2	0.447
0.02	0.141
0.002	0.0447
0.0002	0.014

For X > 1									
No. of whole digits in value X	1	2	3	4	5	6	7	8	Continue the pattern of even then odd values
Which Side (L or R)?	L	R	L	R	L	R	L	R	Continue pattern R,L,etc
No. of Digits in the answer $\sqrt{}$	1	1	2	2	3	3	4	4	Continue Pattern 5,5,etc

Number to have a square root taken	Answer on slide rule
2	1.41
20	4.47
200	14.1
2,000	44.7
20,000	141.

Using the K Scale (Cubes X³ & Cube Roots ∛X) :

- The K Scale is a triple scale and is the cube of the values on C & D
- That is its range is 1 to 10, 10 to 100, and 100 to 1000.
- Rules for Taking Cubes :
- For the Rules consider the K scale divided into 3 sections from one index to the next.
- The sections are Left, Middle, and Right
- For All Values under consideration for cube root extraction :
- Divide the Number into groups of 3 (starting at the decimal point) and go left or right of the decimal point as needed in creating these groups
- Look at the Left-most group with non-zero digits in it
- If the group has 1 digit then use the Left portion of the K scale (for example 1-10)
- If the group has 2 digits then use the Middle portion of the K scale (for example 10-100)
- If the group has 3 digits then use the Right portion of the K scale (for example 100-1000)
- To continue further, continue counting in 3's.
- An easy way to find where a value falls if <1 :
- Let right-most 1 be 1 and go backwards as powers of 10 for any given decimal value :
- Example 0.00X is in the thousands place
- The right-most '1' is 1000/1000,
- The next '1' left of it is 100/1000,
- The next '1' left of it is 10/1000
- And the left-most is 1/1000
- If there are more restart at right-most and continue (Note the scale wraps around then)
- The number of groups left of the decimal point determines where the <u>decimal point</u> in the <u>answer</u> falls :
- If there are two groups (of 3) left of the decimal point (complete or not !) (means values 1,000 – 999,000) then there are 2 figures left of the decimal point in the answer.
- Values : X,XX0 to XXX,000
- If there is one group of 3 (values 1-999) then the answer has one value left of the decimal point
- Values : X.XX to XXX.
- If there are 3 groups to the right of the decimal point (where one or more falls in the tenths-hundredths-thousandths columns) then the answer has 3 figures right of the decimal point starting just past the decimal point
- Values : 0.XXX to 0.00X XX-,---
- Values : 0.000 XXX to 0.000 00X XX0 ---
- If the number has four groups to the right of the decimal point where the first group is all zeroes and at least one value is in the ten-thousandths,

hundred-thousandths, or millionths place, then the answer will have 4 figures to the right of the decimal point where the first is a zero (0).

Decimal value for X 0 < x < 1		Number of Zeroes in value		
Which portion of K scale to read?	R, C, L	R ,C ,L	R ,C ,L	R ,C ,L
Q number of zeroes in value X	0,1,2	3,4,5	6,7,8	9,10,11
Number of zeroes in answer to $\sqrt{}$	0	1	2	3

whole values for X X > 1		Number of Digits in value		
Which portion of K scale to read?	R, C, L	R, C, L	R, C, L	R, C ,L
Q number of digits in value X	1,2,3	4,5,6	7,8,9	10,11,12
Number of whole digits in answer to $\sqrt{}$	1	2	3	4

Using the L Scale (Log Base 10 (N)) :

- The L Scale is the Log value of the C & D scales
- It is effectively used for determining powers and roots for a wide range of values. Note that the powers do not have to be whole numbers as well as roots can be any fractional value one is interested in determining
- The L Scale provides the log of values 1 to 10 with no changes
- If the Number is <1 (includes <0) or >10, the log value has both the Characteristic (the value before the decimal point) and the Mantissa (the value after the decimal point found on the scale)
- Note that the mantissa will be the same for a given value independent of the decimal point.

-
- For example the log (2) mantissa is 301, as is for log (20), log (200), etc – the only difference is the Characteristic, so log(2) = 0.301, log(20)=1.301, log(200) = 2.301, et al
- When the number is <1, can use scientific notation to determine the value to add to the log of the number in question. Log(0.2) = Log(2 x 10^{-1}) = 0.301-1 = -0.699
-
- **Procedure below for logs in general :**
- First Look up the Log value for any given number treating it in Scientific Notation format so that it is >1 and <10
- Locate the characteristic of the scientific notation value on the C or D Scale and look below to the L Scale
- This L value is the Mantissa
- Add the Mantissa to the Exponent of the Scientific Notation exponent
- This Sum is the Final Answer
- The L Scale is useful for X^N and $X^{1/N}$
- To solve – Look up X on C and find its log value on L
- Then Multiply or Divide as needed the N or 1/N value involved in the problem using the rules for multiplication or division. Be sure to watch the decimal point.
- This new value, Q, is computed
- Now Search for the Q Mantissa on the L scale
- Note that if there is a Characteristic, it becomes the power of 10 for decimal placement of the answer!
- Alternative to finding a log for values between 0 and 1
- If a desired value for log(X) has 0<X<1 as a decimal in the tenths place, the C1 scale can be used. Reading 2 on C1 the log value is the log of (0.2) for example.

Using the S Scale (sin(θ) & sin^{-1}(θ)) :

- The Sine (S) Scale has values representing the angles from approximately 5.5° to 90°
- When the cursor is on any value of the S scale (the left number printed in black typically) its sine value can be found under the cursor on the C or D Scale
- Note : Since Sine functions range from 0 to 1, all values read from C or D are decimal and begin as 0.XXX (for most models)
- Note : Some Box-style Post Slide Rules have a cursor on the backside where the S scale often is and with a value on it the sine & tangent (T scale) of the angle is found on the B scale!
- For angles 0° to 5.5° there is often an ST scale and represents both Sine and Tangent functions as these have approximately the same values for small angles such as these (also the decimal is 0.0XXX)

-
- Many Sine Scales have a second set of numbers written in red to the right of the values on the scale –
- These are the complimentary angles hence they represent the Cosine of those angles listed in red
- Recall $\sin(\Theta) = \cos(90° - \Theta)$
- To multiply by the Sine of a given angle simply place the angle over the appropriate D scale index
- Next read along D to the value to be multiplied by
- Find the Answer on the C Scale above that point on the D scale
- Remember to watch the decimal points since the C value is a decimal value
- To divide by the sine of an angle simply place the angle on S scale acting as the dividend over the divisor on the D scale
- Find the Answer at the C scale corresponding D scale Index
- If the square to the sine of an angle is needed since the sine value is on C, the square of the sine value is found on A.
- To find the log of an angle for a trigonometric function such as sine, look up the sine of the angle on C scale, then read on the corresponding L scale for the log of the sine of this angle.

Using the T Scale (tan(θ) & tan⁻¹(θ)) :

- The Tangent (T) Scale has values printed in black from 5.5° to 45° when read left to right
- The values (typically red) read right to left are 45° to 84.5°
- Like the directions for Sine (S) Scale,
- To read the tangent (Θ)
- place the cursor on the desired angle
- If the angle is from 5.5° to 45°, the Tangent to this angle is found on the C or D scale and is read as a decimal 0.XXX
- Recall that $\tan(45°) = 1.00$
- In fact this value lines up with the Right Index of C & D
- If the angle is 45° to 84.5° read the T Scale from Right to left to find the value and place the cursor there
- Answer to the Tangent of this angle is found on the C1 Scale
- The values on the C1 scale are read as whole numbers as the line appears
- (Note there are slide rules that have two T lines, hence are read from left to right on the C scale and one must keep in mind that the C values start at 1 (tan 45°)
- Like the S scale since the tangent (like the sine value) value is on the C scale the rules for multiplying and dividing are the same as noted in the S Scale section
- Note since $\tan(\Theta) = 1/\cot(\Theta)$ one can compute these as needed

Using the ST Scales :

- If your Slide Rule has a ST Scale it can be used for small angles for both the sine and tangent function as these are very close to each other in value between $0.6°$ and $5.5°$
- Much like S Scale and T Scale readings the angle is on the ST scale and the reading comes from the C Scale.
- Important Reading Scale Note : Here however the readings range from 0.01 and 0.1, so the reading has 10^{-2} as its exponent.
- What if there is no ST Scale :
- In approximations :
- We can use : $\sin(x) = \tan(x) = \dfrac{x}{\frac{180}{\pi}} = \dfrac{x}{57.3}$
- Note that this approximation also is the radian value for that angle!
- Here x is the small angle and using the C and D scales then provides a reasonable value for the sine or tangent of the small angle (within the same range).

Using the Log (LLn Scales) :

- The LLn Scales are used to raise a number in question to a power or find a root of that number.
- Each line is a power of e (2.718...), some models it is the power of 10, of the next line in the list (LL1, LL2, et al).
- That is to say each line is e to some power (from -10X (LLO3) to +10X (LL3))
- The Table of Names and Powers for LL scales are below showing both the power and the range of values to be found on those scales. :
- Looking at the tables, it is clear that the spacing needs to be examined carefully when reading the scales. Be sure to look at the two primary values that your values is between (as an example) and then look at the number of secondary divisions to determine the appropriate value.
- These scales are indispensible when looking for any given power or root as needed (not necessarily whole number ones either).
- Also since each is 10x the line before it, for example LL3 is 10x LL2, and LL2 is 10x LL1, and so on, hence LL3 is 100x LL1. This means that for any power of 'e' which is used for the scale when reading from the C or D scale each is 10x the others.
- For example e^2 is found by looking at the D line for 2 and reading from the e^{1-10x} line, which is LL3 and is approximately 7.4
- Better still, what if you wanted the inverse of e^2 or 7.4? Simple read its value on LL03 – Be careful in reading the scales, these are in reverse first off and have decimal values – so read it from right to left
- Here the inverse of 7.4 is between 0.1 and 0.2 and reads 0.135 – try this yourself and practice reading the scale properly.

-
- What about $e^{0.2}$, that is found on the LL2 scale, (yields a value nearly 1.222)
- and $e^{0.02}$ is on the same cursor line with the cursor the whole time on 2 on D.
- This is one of the great values of the LLn scales. Any value from 1.001 to nearly 100,000 can be used from LL0 to LL3 and its inverse is present as well if needed. That would be 10^{-5} to 0.999 found on LL00 to LL03. It is typically written in this range though, due to the effectiveness of the slide rule : having a range of values (up to 80 inches in length on these LLN scales) of running from 0.00005 to 20,000.
- Looking at a value on say LL0 scale and reading the value below on LL1 scale it becomes that number raised to the power of e 10x more. So an increase in number is 10x the prior line.
- The LL0 value if read with the LL2 scale will then be that value raised to the power of 100.
- Going from a higher LL scale number to a lower one means that it is $1/10^{th}$ power per each jump.
- To raise to a negative (−) exponential power : This can be done directly or find the positive value then its inverse on the corresponding negative exponent line.
- If raising negative exponent to positive exponent values :
- If one wants to find 10x the value in power simply read from one line to the next, such as going from LL1 to LL2 in essence multiplies the power by 10 if needed.
- Going in the reverse direction divides the power by 10, of course.
- For example read a value on −LL1 scale and find that value to the 10th power on −LL2 scale.
-
- Note : the −LL Scales are normally written in red. And the values increase from right to left (as do all other inverse scales)
- Recognize that −LLX scales are the inverse of LLX scales correspondingly. So the inverse of the values found on LL1 are found on LL01 also known as −LL1.
-
- **Using the LLn Scales in general :**
-
- To find X^N :
- Look up the value X on the LL Scale.
- Slide the Index of the C (1) scale over it.
- Move the cursor along the C Scale to the power (N) and
- Read below the answer on the needed LLX scale
- For example, 2 is squared, 3 is cubed, 5 is the 5th power – but you can do any power in between too – such as 4.7^{th} power for whatever reason is needed :
- (be sure to estimate since it may have gone from one of the LLX Scales to the next one in line)

43

-
- **One could find roots as well :**
- For a given value, X find this on the needed LLX scale
- Place the needed root over it (N) on the C scale
- Note : Recognizing the this is read as 1/N mentally
- Now read in conjunction with the C scale Index the value on the appropriate LLX scale the root value
- Essentially taking the root is the reverse of the process of determining a power – much like doing multiplication and division with the slide rule in terms of directions.
- Note : this process works for –LL Scales as well.
- LL Scales are not good for numbers very close to 1, such as 1.001 or 0.999.
- There is an approximation for this value for small values of 'n' with : $(1 + n)^p = 1 + d^p$

Name	Power	Range	Name	Power	Range
(+)LL3	$e^{+1.0x}$	e to 22k	(-)LL03	$e^{-1.0x}$	1/e to 1/22k
(+)LL2	$e^{+0.1x}$	1.1 to e	(-)LL02	$e^{-0.1x}$	0.91 to 1/e
(+)LL1	$e^{+0.001x}$	1.01 to 1.1	(-)LL01	$e^{-0.001x}$	0.990 to 0.91
(+)LL0	$e^{+0.0001x}$	1.001 to 1.01	(-)LL00	$e^{-0.0001x}$	0.999 to 0.990

Note : in table, the + and – denotation is used together, such as –LLN and +LLN, on some slide rules, while on others it is LLN and LL0N denotation depicted

Tricks & Tips in Slide Rule Use :

- Check alignment of scales and adjust as needed
- Always be sure of the level of precision of the scales for proper reading. Most 10" Slide Rules are quite good to 2 significant digits with even a 3rd estimated significant digit being possible.
- In terms of all calculations undertaken, Be sure to first estimate the answers ahead of time mentally
- Often it is a good idea to convert most values to scientific notion both for ease of calculation and determination of decimal placement.
-
- **Summary of Scientific Notation Rules :**
- **The Rules for Decimal Placement :**
- 1) Always first estimate the Answer.
- 2) Convert all values into Scientific Notation.
- 3) For Any Multiplied Values, Add up the Exponents and For Any Divided Values, Subtract the Exponents.
- 4) If for any single Multiplication operation the Slide moves to the Left, then add +1 to the exponent total (Why? Because in essence, we have gone off this scale and are adding two values that extend beyond the scale in front of us to a next one in line, which is 10x the line before it)
- 5) If for any single Division operation and the Slide moves to the Right, then -1 from the exponent total (Why? Because in essence, we have gone off this scale and are subtracting two values that place the answer on the scale to the left of the one in front of us, which is 1/10th the line before it)
- 6) Mentally treat any two values independently (just as whole numbers now) but keep track of what the exponent will be through the rules.
- 7) That is to say answer the question as to where the Slide has moved and factor in that addition or subtraction as needed for each operation independently.
- 8) Take this final answer and use the exponent figure arrived at to determine the decimal placement!

- **Alternative Decimal Place Method :**
- **Summary of Decimal Rules using the Counting Digits Method found in some slide rule books :**
- The key to the digits method is to tally the number of digits in a number.
- The basic rule for this is this :
- For numbers X > 1 the number of digits is simply the number of digits in the number.
- For example : 7 has 1 digit, 70 has 2 digits, and 70,000 has 5 digits, etc.
- What if the number is greater than 1 but has decimals, though?
- For example the number is 23.45, how many digits does it have?
- It has only 2.
- In the case of X > 1, all decimal values are overlooked in the digit count.
- So the next question then is, what of values 0 < X < 1 then, what is their digit count?

-
- In the Digit Count Method all values 0.1 to 0.9 expressed as nothing more than 1/10ths values have 0 (zero) digits.
- So 0.6 has 0 digits, for example.
- With each decimal place it is like a reverse number line :
- 0.01 has -1 number of digits,
- 0.001 has -2 number of digits,
- 0.0001 has -3 digits,
- And so on...
- To help remember this the value of digits is essentially the negative number of zeroes past the decimal point.
- Summary of these ideas on Counting :
- For numbers greater than 1, count all the numbers up to the decimal and treat these values as positive. For numbers less than 1, count the number of places up to the first nonzero number and view these as negative values. Sum up these numbers in multiplication and when the slide extends to the right, subtract one from the sum.

-
- With the number of digits in all the values you have in your problem now look to the following rules when it comes to multiplication and division :

-
- Multiplication with C & D Scales :
- 1) If the slide projects to the right of the stock during multiplication, the digit count for the product is one less than the sum of the digit counts for both the values in your calculation (called the multiplicand and the multiplier).
- 2) If the slide projects to the left of the stock, the digit count for the product is equal to the sum of the digit counts for the multiplicand and the multiplier.
- Division with the C & D Scales :
- 1) If the slide projects to the right of the slide rule stock during a division, the digit count for the quotient is one more than the digit count for the dividend (the numerator) minus the digit count for the divisor (the denominator).
- 2) If the slide projects to the left of the stock, the digit count for the quotient is equal to the digit count for the dividend minus the digit count for the divisor.
- Multiplication with the C1 & D Scales :
- 1) If the slide projects to the right of the stock during multiplication, the digit count for the product is equal to the sum of the digit counts for the multiplicand and the multiplier.
- 2) If the slide projects to the left of the stock, the digit count for the product is one less than the sum of the digit counts for the multiplicand and multiplier.
- Division with the C1 & D Scales :
- 1) If the slide projects to the right of the stock during division, the digit count for the quotient is equal to the digit count for the dividend minus the digit count for the divisor.
- 2) If the slide projects to the left of the stock during division, the digit count for the quotient is one more than the digit count for the dividend minus the digit count for the divisor.

-
- <u>Other Ideas :</u>
-
- Examine the problem carefully, working from the inside out, and use the best scale for that calculation
- Recognize that values can be found easily with given scales – squaring, square rooting, cubing, cube roots, sine values, etc. When moving the cursor its alignment may find it at a place to make use of these as needed.
- Proportions (which also conversions and 3-variable functions can be treated as) are very straightforward and easy to set up and solve on a slide rule.
- Note that most things are Ratios or Proportions and the Slide Rule is the best tool for these calculations since it sets up all similar ratios instantly with any setting!
- Create a needed scale! Take the activity where we used the C & D scales with the C1 scale. If there is no C1 scale, simply invert the slide in the slide rule and use the reversed C scale as if it were a C1 scale now.

- **<u>How to Add Numbers on a Slide Rule :</u>**
- Suppose we have two numbers X & Y so that we want their sum (X + Y) and yet use a slide rule!
- Let's first rearrange this expression : $X*(1 + \frac{Y}{X})$
- With a slide rule simply set the Y value on C over the X value on D.
-
- If Y > X read answer above the D index from the C scale as whole number with decimal and add one to it. Note that the decimal placement rules also must apply however.
- OR
- If Y< X read answer at D index on the C scale as decimal value and add one so that it is 1.***. Note that the decimal placement rules must also apply however.
-
- Now Take the result and find it on D and place this value under the left C index.
- Now, Regardless of Y > X or Y < X, now read along the C scale to the X value and find the answer for the sum of X + Y on the D scale below! Be certain to have an estimate of your answer and employ the proper decimal placement rules for reading the slide rule.

- **Other Things to find on the Slide Rule :**

- There are many ratios that can give good estimates in situations :

- Set up these ratios and find the known value and opposite it is the sought after answer -

$$\frac{\text{Circumference of Circle}}{\text{Diameter of Circle}} = \frac{355}{113} \qquad \frac{\text{Feet}}{\text{Meters}} = \frac{82}{25} \qquad \frac{\text{Atmospheres}}{\text{Feet of Water}} = \frac{23}{780}$$

$$\frac{\text{U.S Gallon}}{\text{Cubic inches}} = \frac{1}{231} \qquad \frac{\text{Feet of Water}}{\text{Pounds per sq inch}} = \frac{60}{26} \qquad \frac{\text{Side of a Square}}{\text{Diagonal of a Square}} = \frac{70}{99}$$

$$\frac{\text{US gallons}}{\text{Liters}} = \frac{14}{53} \qquad \frac{\text{US Gallons}}{\text{Imperial gallons}} = \frac{6}{5} \qquad \frac{\text{Inches of Mercury}}{\text{Feet of Water}} = \frac{15}{17}$$

$$\frac{\text{Yards per Minute}}{\text{Miles per Hour}} = \frac{88}{3} \qquad \frac{\text{Pounds per sq yard}}{\text{Kgs per sq meter}} = \frac{46}{25} \qquad \frac{\text{Weight of fresh water}}{\text{Weight of sea water}} = \frac{38}{39}$$

$$\frac{\text{US Gallons of Water}}{\text{Weight in pounds}} = \frac{3}{25} \qquad \frac{\text{Ounces}}{\text{Grams}} = \frac{6}{170} \qquad \frac{\text{Diameter of Circle}}{\text{Side of equal square}} = \frac{79}{70}$$

$$\frac{\text{Inches}}{\text{Centimeters}} = \frac{26}{66} \qquad \frac{\text{Pounds}}{\text{Kilograms}} = \frac{75}{34} \qquad \frac{\text{Area of Circle}}{\text{Area of Inscribed Square}} = \frac{322}{205}$$

Ch.IV
Gauge Marks and Scales of the Slide Rule

The Scale

The primary use of the slide rule is seen in our activities for multiplication and division. This was done by using non-linearly divided scales that are divided instead in a logarithmic manner (that is to say the numbers were not evenly spaced like on a ruler but here by logarithmic distances).

Why use the logarithmic spacing of the numbers ? The most basic linear form utilizes two scales where the numbers are logarithmically spaced. Due to the mathematical properties of logarithms, the spacing of the numbers on these lines allows for easy and rapid multiplication and division with the same set of rules despite the type of slide rule used. (This idea is explored in the history section in more detail discussing the discovery and importance of logarithms to history which the slide rule can be seen as the physical visual manifestation of this math form).

Many slide rules have more than 2 scales, and these can be used for many other mathematical operations as we have seen, such as : squaring, square roots, cubing cube roots, raise to various powers, taking a various root of a number, common and natural logs, and trigonometric values such as sine, cosine, and tangent.

Standard scales spaced in a logarithmic fashion, like C and D are said to be **Single Scales**. That is, there is one run from 1 to 10 in the distance for the slide rule, such as 25 cm (i.e. 10 in). If there are two scales running from 1 to 10 in the same distance, it is said to be a **Double Scale**, such as for A or B scales. Of course if there are 3 runs of 1 to 10, which is the case for the K scale, this is a **Triple Scale**.

Another type of scale is the **Folded Scale**. Instead of starting at the Index point of 1, it begins at or near another point on the logarithmic line. The usual choice is at pi (π). The choice of pi was two-fold. One it allowed for using pi as an Index, so that calculations involving circles and cylinders could easily be done and two, by cutting the original C or D scale at a different point, made alignment more convenient for times when multiplication involved numbers from each end of the regular fundamental scale.

The C and D scale are the primary or fundamental scales of the Slide Rule and all others are based on them. For example, we see from the activities, the tangent (as well as the sine) scale is read in relation to the C or D scale. This idea is further illustrated in the table which lists in alphabetical order the scales, their mathematical relationship to C or D and a brief description of that scale.

Table of Scales and Uses

SCALE	Mathematical Relation to C or D	DESCRIPTION and Other Notes
A	X^2	Square Values of Fundamental D scale. Double scale, Opposite to B scale
A1	$1/x^2$	Reciprocal of A scale, Reciprocal of square of D scale
B	X^2	Square Values of Fundamental D scale. Double scale, Opposite to A scale
C	X	Fundamental (Single) Scale. On Slide opposite to D scale
CF	πx	Folded Fundamental Scale, starts at π. Opposite to **DF scale**
CI	$1/x$	Reciprocal of Fundamental Scale. On Slide – Basically a Reverse Scale
CIF	$1/\pi x$	Reciprocal of CF scale. Reciprocal of Folded Fundamental Scale C
D	X	Fundamental (Single) Scale. On the Stock opposite to C scale
E	e^x	Log-log scale – see LL3
K	X^3	Cube Values of the Fundamental D scale. Triple Scale
L	$\text{Log}_{10}x$	Mantissa of the common logarithm of Fundamental D scale value
LL	$\text{Ln}(x)$	Mantissa of the natural logarithm of the Fundamental D scale value
LL0	$e^{0.001x}$	Scale yields 'e' raised to 0.0001*x power, where x is read from the fundamental scale. Positive Log-log scales are used to raise a number to some exponent or find roots >1. LL scales are to enable one to raise values to a power and take roots very readily
LL1	$e^{0.01x}$	Scale yields 'e' raised to 0.01*x power, where x is read from the fundamental scale.
LL2	$e^{0.1x}$	Scale yields 'e' raised to 0.1*x power, where x is read from the fundamental scale.
LL3	e^x	Scale yields 'e' raised to x power, where x is read from the fundamental scale.
LL00	$e^{-0.001x}$	Scale yields 'e' raised to (-0.001*x) power, where x is read from the fundamental scale. Negative power means $1/e^{-0.001x}$. Negative exponent log-log scales are used to raise numbers to a power or find roots <1
LL01	$e^{-0.01x}$	Scale yields 'e' raised to (-0.01*x) power, where x is

		read from the fundamental scale.
		Negative power means $1/e^{-0.01x}$
LL02	$e^{-0.1x}$	Scale yields 'e' raised to (-0.1*x) power, where x is read from the fundamental scale. Negative power means $1/e^{-0.1x}$
LL03	e^{-x}	Scale yields 'e' raised to (-x) power, where x is read from the fundamental scale. Negative power means $1/e^{-x}$
P	$(1-(0.1x)^2)^{1/2}$	Pythagorean Scale. Cosine of $\sin^{-1}$ D scale
R1	$\sqrt{}$	Square root of Scale D value. Scale twice the length of D scale. R1 runs 1 to 3.2, R2 runs 3 to 10. Aka: W1, W2 and Sq1, Sq2
S	$\sin^{-1}x$	**Scale D** value is the sine of the angle in degrees read on the **S scale**. Runs 5.7° to 90°
ST	$\sin x$, $\tan x$	Sine & Tangent of small angles 0.58° to 5.73°. Same and S&T scale. Used since sine and tangent are similar at these angles
T	$\tan^{-1}x$	**Scale D or C and CI** value is the tangent of the angle in degrees read on the **T scale**. Runs 5.7° to 84.3°. D scale read for angles 5.7° to 45° increasing. CI scale read for angles 45° to 84.3° (printed in red and in reverse order)

The table above is by no means all of the scales, nor does it trace the differences in names for the same scales as they changed through time. It is a representation of the most common scales and represents a list that most slide rules commonly have a subset of.

As noted earlier, one could simply use a 9 scale form (such as in the activities) and be able to compute the overwhelming majority of problem types that even a regular scientific calculator is capable of today.

This does not mean that the other scales were merely for show, though there was probably some pride and showing off in the office if one had a more advanced model I imagine. The other scales were applicable to various disciplines, such as chemistry, physics, electrical engineering, and the like.

This idea is particularly true when one considers not just the right scale combination form, but also relevant and useful gauge marks that may be on the scales. Some are listed and described below in the accompanying table. Like the scale table, this is not complete and represents a cross-sectional view of them.

Table of Gauge Marks on the Slide Rule

Besides the numbers on a slide rule perhaps you have noticed some out of place marks or letters? These are gauge marks. These marks are at specific places corresponding with their value and have a given prescribed mathematical value to the user of the slide rule.

The table below lists some of the more common ones. This table is very far from complete but illustrates some of the possibilities. Note that not all makers use the same letters nor do they have all of the same marks on their various models. Some were probably put on there since those models may have been marketed to a particular set of professions where that mark would have value and use.

Look at some of the examples such as symbols for the weight of copper conductors, watts in one horsepower which clearly had applications in the electrical industry. Other constants as gauge marks can include the acceleration due to gravity for calculations in physics.

More common ones include the conversions of radians to degrees and vice versa as angular measures in calculations were commonplace.

Probably as no surprise the most common gauge point is π, since it can be used for all circular calculations. Using π we realize the value in this. Knowing where it is at allows quick calculations of either multiplication or division as needed.

Gauge Marks	Meaning	Value
C	Square root of $4/\pi$ Circle area calculations	1.128
G	Acceleration due to Gravity (metric)	9.8 m/s^2
G	Acceleration due to Gravity (English)	32.2 ft/s^2
L	Natural log of 10 Convert $\log_{10}$ & $\log_e$	2.3026
R, p, or r	$180/\pi$ Convert radians & Degrees	57.3
p′	Minutes in a radian	3438
Q	Radians in one Degree	0.01745
	$\pi/4$	0.7854
π	Pi – ratio of circle Circumference to Diameter	3.1416
W	Weight of copper	111000
-	Watts in one Horsepower (hp)	745.47 (746)

There are also common conversion marks and value marks with no special symbols on various slide rules, such as on the Fowler's Circular Calculators. Many of them have value marks for square-root of 2, square-root of 3, along with pi, and conversions for inches to centimeters, kilograms to pounds, square centimeters to square inches, and the like.

Ch V
The Story of Waves

Primary Reading about Wave Types and Wave Components :

A **Wave** is any disturbance that repeats regularly in space and time which is transmitted from one place to the next with no actual transport of matter. Waves transport energy and are used to transport information. For example, drop a pebble into a pond to create a set of waves. They radiate out from the impact spot of the stone on the surface along the surface of the water. The gravitational potential energy of the stone you dropped turned into kinetic energy which in turn became the sound, heat, and wave energy of the pond. The waves move out and can do work on something they encounter, such as erode the shoreline boundary. In the area of information, light waves are used to tell us about the stars in the universe (their temperature and composition, for example) and radio waves are used to transmit information.

There are 2 basic types of waves : **Mechanical and Electromagnetic** (which will be discussed) and each of these have component parts that have familiar terms.

Waves occur in nature with a great deal of regularity and we make use of them and generate them too for use. Waves on the ocean as noted at the shoreline are common to us all. Any sound is a wave. This includes those people can hear and those they cannot (infrasonic and ultrasonic). Sound waves, a mechanical wave, can travel through the Earth, such as when a earthquake occurs and is recorded even on the other side of the world. Light, an electromagnet wave, is a wave as are all of the other forms of radiation, such as radio waves, infrared radiation, ultraviolet radiation, x-rays, and gamma rays.

The classical description of a wave in the form of a picture shows a vibration pattern that oscillates with regularity with time on an x-y axis where x is time and y is displacement. The form is a sinusoidal wave typically (though there are other types).

One of the first parts of a wave to take notice of is what is called the **Wavelength (λ)**. The wavelength is the distance from one point on a given wave to its corresponding point on the next wave in succession. This can be from crest to crest (the high point of the wave) or trough to trough (the low point of the wave) and any other chosen point. The unit for wavelength is a distance unit and is typically meters (m).

The height of the wave has a name as well. **Amplitude (A)** is the distance from the midpoint of the wave to its crest (or trough). It is measured in distance units, like meters (m). The amplitude is related to the energy of a wave. For a Mechanical Wave, the energy of the wave is proportional to the square of the Amplitude. For example, this would be related to the Intensity of Sound, which is a measure of its energy and it is a mechanical wave.

Since a wave is something that occurs with regularity, there must be a measure of this behavior. Frequency is this measure. **Frequency (f)** is the number of events (cycles, vibrations, oscillations, or any repeating event) per unit time. This measure is how many vibrations occur in a second. A frequency of 20 is twenty times per second while 100 is one hundred times per second. This measure has the unit Hertz which is $\frac{1}{seconds}$ (Hz or s^{-1}).

If we know how many waves per second are occurring, then the inverse of this tells us the amount of time for one wave. **Period (T)** is the amount of time to complete one cycle. It is measured in time units, typically seconds (s).

The relationship of period and frequency can be expressed mathematically therefore. They are each the inverse of the other. Their product for a given wave is equal to one.

$$\textbf{frequency} = \frac{1}{\textbf{Period}} \qquad\qquad \textbf{f} = \frac{1}{\textbf{T}}$$

The other key variable is the **speed of the wave**. It has the same definition as speed for any physical object, though here it is a wave. Wave Speed is the distance covered by the wave per unit time.

$$\textbf{v} = \frac{\textbf{d}}{\textbf{t}}$$

The distance one wave travels (wavelength λ) is in the period of the wave (T), so we can write the regular equation as :

$$\textbf{v} = \frac{\lambda}{\textbf{T}}$$

But the Period is the inverse of frequency (f)
so we can rewrite the equation as :

$$\textbf{v} = \lambda\textbf{*f}$$

These last two equations are known as the **Wave Equation**. Knowing the wavelength and the period and/or the frequency of a given wave, we can determine its speed.

Regardless of type of wave, all of these equations can be used for describing them. Notice the similarity to kinematic waves in describing motion.

The first major category of waves are **Mechanical Waves**. Mechanical Waves are waves that require a medium. A medium is a physical material that the wave or disturbance travels through. Mechanical Waves primarily include Sound Waves.

The sound waves we are familiar with are the ones in air that we can hear. These range in frequency from 20 Hz to 20,000 Hz. Despite this wide range of hearing capacity, there are still greater ranges. The **infrasonic** range is for those frequencies less than 20 Hz (infrasonic means below the range of hearing). Some animals, such as elephants have been found to transmit in this range. The frequencies above 20,000 Hz are called **ultrasonic** (meaning above the hearing sound range). These frequencies are used by bats, dolphins and others that use **sonar** (sound navigation ranging). We use them presently in medicine to create internal pictures of our bodies and other objects.

An important idea to note here is the necessity of a **medium**, such as air for sound. Imagine you place an electric battery-powered alarm clock in a bell jar and it goes off. Through the glass it is still audible. Now use a vacuum pump and draw out as much of the air as is possible (one cannot get all of it). What is noticed is that the sound seems to fade to nothing at all. If the alarm clock were the classic one where a striker is moving to and fro and hitting two bells, we see this still happening through the glass of the jar, but now there is no sound. If we could touch the clock we would feel the vibration, but now it has nothing to travel through, hence no sound!

This is similar in reasoning to why a person who has taken in some helium sounds so different (much higher pitch) than before when speaking in air. Helium is a gas, much like air is, but it has a much lower molecular weight than air. (Helium is the second lightest element, while air is composed of diatomic forms primarily of nitrogen and oxygen). The lower molecular weight of the gas allows our vocal chords to vibrate faster, hence at a higher frequency or pitch. So we have a 'squeaky' sounding voice with some helium passing over our vibrating vocal chords. **(Note : Do not do this as there may be more than helium in the helium-filled balloon, and too much of non-oxygenated gas can be harmful).**

Sound waves are not just sound in air, though, since sound can travel through liquids and solids just like gases. The most well known examples are directly from nature : earthquakes and the sonar and songs used by dolphins and whales. In the case of earthquakes the colliding plates of the Earth which can give way from time to time due to the rocks breaking sends a wave of energy (sound) through the solid rock itself. Seismographs around the world record these events and with geometry the location and intensity of the earthquake can be determined.

What is more important about earthquake waves is that the entire Earth interior is not solid and part of it is liquid! We cannot venture there and for many years the interior of the Earth was a total mystery Earthquake waves are not of just one variety, but 3. The first type we may be familiar with are **Surface Waves**, as these are on the surface and effect the local surroundings of where the earthquake occurs. The next are the **Primary Waves** (aka **P Waves**). These are **Longitudinal Waves** (aka Compression Waves) – which are waves that vibrate along the direction of travel. These have regions of 'compression' and 'rarefaction' (higher and lower pressure and density regions). These can travel through both solids and liquids, hence can go through the Earth. The other notable wave is the **Shear Wave** (or **S Wave**) which is a Transverse Wave form. **Transverse Waves** are ones that vibrate at right angles to the direction of travel for the wave. These latter waves cannot travel through a liquid, hence there is a 'shadow zone' on the opposite side of the Earth from an earthquake where these waves do not show up as they outline the outer core region which is liquid.

In a solid the compression waves speed is determined by the medium's compressibility and density, while the speed of the shear waves in solids is determined by the material's stiffness, compressibility, and density.

An interesting fact is that waves in solids are faster than in liquids and these are faster than waves in gases (on the average). The speed of sound in air is about 331.5 m/s, while in helium it is 972 m/s. The speed of sound in water is about 1490 m/s yet in sea water (since there is salt and other minerals dissolved in it) it is 1530 m/s. The speed of sound in solid copper is 3560 m/s, while in solid aluminum it is 5100 m/s and in solid iron it is 5130 m/s. (Note that this is primarily true for solids that are crystalline in nature). It is not the density of the material that accounts for the differences, but is, instead, due to the stiffness and compressibility of the materials involved. The easier a material is to compress, the slower the speed. Solids are very hard to compress, hence a faster speed as compared to gases which are readily compressed so have a slower speed.

Noting the wave equation, the wavelength and the frequency are related directly to wave speed. The first thought is that since wavelength is part of the equation, then speed is determined by the wavelength of the wave. This is NOT the case! Imagine that a person is sitting 10 m away and is going to play the high pitch piccolo while 10 m in an opposite direction is a person going to play a low pitch tuba. With a signal by a light being turned on, they each simultaneously play a note. Which arrives at your first? Neither. They both arrive at the same time since they are not just equidistant, but they travel at the same speed. Interestingly, this fact is true for both Mechanical and Electromagnetic Waves. Though, each wave type does not travel at the same speed they have the same properties within their own realm. All sound waves in air at the same temperature and other atmospheric conditions, will travel at the same rate of speed (about 331.5 m/s). The speed will vary with temperature and yet be constant for a set of frequencies. This same idea as an experiment would show the same situation for light of different colors, for example blue and red. Of course the light would travel at the much faster speed of the speed of light (3×10^8 m/s).

This brings us briefly to the other type of wave, namely : **Electromagnetic Waves**. These are waves that do not require a medium and are transverse wave forms composed of vibrating electric and magnetic fields that are at right angles to each other and to the direction of travel. Light is the most well known electromagnetic wave and there is a whole family of them : radio waves, microwaves, infrared waves, ultraviolet rays, x-rays, and gamma rays. Though they each have their own unique wavelengths (and hence frequencies) they all travel at the same speed in a vacuum, namely what is called the speed of light. This is a cornerstone of Einstein's famous laws of Relativity (Special Relativity in particular). Regardless of the relative motion of the observer, all parties will agree that the speed of light is constant! Also it stipulates that the maximum speed in the universe is this value and that matter cannot go this speed, only energy of these waves (and gravitational waves if found to exist) can.

The consistency of the speed both for sound and of light is important to all of science, particularly in the form of the **Doppler Effect**. This comes from Austrian scientist Christian Doppler (1803 – 1853) who noted the apparent change in frequency of a passing train from the perspective of an observer at a station. As a train approaches and it sounds its whistle, the whistle's frequency is higher than if it were a stationary sounding whistle. Once passed, and the moving train again sounds its whistle, the frequency is lower than its normal frequency. Note that in all cases, stationary, approaching, or receding, the wave speed of the waves (for the whistle) are constant. Why then does the frequency change? As the train approaches, the sound wave crests from the whistle are closer together when they are encountered by the stationary observer at the station. Hence they are encountered more often than the stationary whistle, so a higher frequency or pitch in this case. When the train recedes (moves away) the wave crests are more spread out, so a lower rate of encountering the wave fronts, so a lower frequency hence lower pitch. Note that this is not the intensity of the wave. A far away object will sound less intense than one close by since the wave spreads out by the inverse-square law for distance (see Inverse-Square Law of Light Activity #11).

The Doppler Effect also applies to car horns and sirens as well. Today the Doppler Effect is used by bouncing radar beams off of moving objects, such as ice particles or water droplets in the air, baseballs, and cars to determine their speeds. In the former cases noted, this is for weather to find whether a system has precipitation, its size, speed, and other important factors about it. In sports, the speed of a ball or race car can be readily found. For law enforcement, the radar gun is used for determination of speed for cars on the roads.

Even more importantly for Light is the Doppler Effect. Any moving light source will give away its relative motion as compared to an observer by a frequency shift in its characteristic wavelengths. This is where we find the terms 'blue-shifted' and 'red-shifted'. If the naturally occurring waves are shifted towards the blue end of the spectrum, the object is said to be 'blue-shifted' and hence moving towards the observer. If the waves are found to be moved towards the red end of the spectrum it is then called 'red-shifted' and moving away from the observer. Note that it does not mean towards the red or blue end of the spectrum itself. Take for example radio waves. They too can be Doppler Shifted, as can all other forms of electromagnetic energy. To move away means they are shifted to a lower frequency than their natural frequency at a stationary mode and to move towards means they are shifted to a higher frequency from their stationary frequency. Not only is this useful for moving objects, but also rotating one too. The side moving towards the observer will be blue-shifted while the side rotating away is red-shifted. This information is also used to determine the relative speeds of objects in the Universe and was used by Hubble to find that the Universe itself is expanding when he was looking in the 1920s through the Hale telescope at distant galaxies.

Color Addition & Subtraction

Sir Isaac Newton used a prism to separate white light into its component parts, what we commonly call the spectrum or a rainbow. He tested each of the colors with other prisms to find they could not be further separated into still other parts, so he concluded they must be the fundamental parts of light itself.
A question might arise, however, what if we were to mix some of these wavelengths of light in the spectrum of light? This is the process of color mixing by addition. Our eyes work on this principle – we have parts of the eyes that see red, green, and blue and these are combined. When red, green, and blue light are added in equal portions, it creates white light. These three colors are called the additive primary colors.

This color addition, however, is not how the color of an object is generated. We use pigments, such as in paint, in order to color an object. When mixing paints, it grows darker and darker with the addition of different pigments. Eventually with all of the mixing of all the primary pigments, an object would appear black. In physics, a black object is one that absorbs all incoming light and releases none (this is an ideal description). The color of an object due to pigments is achieved through color subtraction (instead of color addition which occurs with light). Each pigment absorbs certain frequencies of light and the others will bounce off (typically they are absorbed and re-emitted). The primary subtractive colors are magenta, cyan, and yellow.

Here is an illustration. Say you have a painted wood block and it is red. It is red since as white light hits it, the red frequencies (plus a little of a couple of others) are reflected off (ie re-emitted) while the others are absorbed. Here the blue and green frequencies are absorbed.

What of the sky – why is it blue? As white light comes from the Sun, the shorter wavelengths are scattered the most. These are the blue frequencies, which are not only scattered, but absorbed by molecules in the air that have the same frequencies as the blue light so it is absorbed and re-emitted in all directions. This is why the blue sky seems to come from all directions – it is all sunlight from the Sun. Since so much of the blue is removed, the Sun has an overall yellow hue to it (do not look into the Sun). When the Sun is at the horizon (rising or setting) since the light must travel through more of the atmosphere more of the blue is scattered than usual leaving the Sun to appear orange to red in color.
This idea that atoms can absorb certain frequencies is a great find. It is the basis of spectroscopy in Astronomy and used in chemistry for atom and molecule identification. Each element will have a set of lines when heated to the point of giving off light which are its characteristic frequencies. With this technique it was found what stars are made of and the fact that the Sun is a star just like all the others in the sky in the 1800s.

Optics :

As we have read, Light and Sound can be described as waves. When studying light waves and their interaction with matter, this is referred to as Optics in Physics.

Waves, and in this case, light waves, interact with matter in a number of ways. Waves can **Diffract** which means to bend or spread out when passing through a narrow opening (as compared to their wavelength the opening is some small multiple of the wavelength). Also, light waves can bounce off of surfaces. This is why there are mirrors and materials that act in a mirror-like fashion (water surfaces). **Reflection**, by definition, is when a wave reaches a boundary between two **media** (the materials the wave encounters), some or all of the wave bounces back into the initial or first medium from where it originated. Light as well as sound can bounce. Bouncing sound is not only interesting when it comes to echoes, but is also used for ultrasound devices as well as the older use of sound as Sonar (sound detection and ranging) where the delay in time of sound echoes are used to detect objects. Unlike sound waves, electromagnetic waves can be used as well, hence Radar (acronym meaning radio detection and ranging) which, by a similar delay in timing method, is used to determine the relative speeds of objects from water droplets in clouds (Doppler Radar) to baseballs pitched to cars on the roadways and planes in the sky.

The concept of Reflection has led to the **Law of Reflection**. It states : for any given wave encountering a surface at some given angle to the normal (Note : The **Normal** is a line perpendicular to the surface - at the point of contact by the wave here) the:

Angle of Incidence = the Angle of Reflection.

This property might seem obvious for a flat mirror, which behaves like a non-spun basketball that is bounce-passed to a friend, but does it hold for curved mirrors (like those in telescopes or stores for security)? Yes it does. Realize here, though, that the normal though 90° to the surface, since the surface is curved will have a direction depended on the curve. In fact this is why the surface of a mirror is curved for a telescope. All of the 'normals', from the curved surface of the mirror, point to a focus. The mirror collects the light and brings it to one place for observation.

What if the light waves pass through a material? This is the behavior called **Refraction**. Defined it is : **the change in direction of a wave as it crosses the boundary between two media in which the wave travels at different speeds.**

Notice a couple of points about the definition. First there needs to be *two distinct media*. It may be very different, such as air and glass, glass and water. These would occur when a ray of light passes through the air, then through the side of a glass aquarium tank and then into the water. Also, the media can be of the same material only with different conditions. For example, Air at two different temperatures will affect the speed of sound differently.

A critical aside is the fact that these topics are referring to Transparent materials as opposed to Opaque materials. Though the definitions are basic, there is more to them than this. The wave does not merely travel through the material uninterrupted. Instead, for example, a light wave encountering the exterior of a transparent glass window is absorbed by the atoms in the glass, which quickly give them off to the next set of atoms and so on as they pass through the glass. The types of atoms have natural frequencies. At ultraviolet ranges of energy the electrons in the glass vibrate. This is because the electrons have natural frequencies that correspond to these waves they encounter. With these ultraviolet frequencies the glass resonates and can hold (hence block) some of these rays. At lower frequencies, such as light waves, these only force vibrations on the atoms and these energies are held for short time periods. What of infrared, then, at still lower frequencies, do they pass through as readily? No. This is because they vibrate the molecules and the molecules of glass have these natural frequencies, hence these energies are held onto and heat the glass up.

What of opaque materials? Instead of the wave energy going through the material, it typically increases the random kinetic energy of the surface electrons, atoms, and molecules of the object. This, in turn, makes the object warmer (much like something being out in the Sun). Some reflect energy better, like metals acting as mirrors, since they have outer loosely held electrons that can more readily re-emit the light that falls on them. Other objects with non-smooth surfaces and this lack of outer looser electrons bounces some of the light (less than shiny metals) and holds back some of it too. This is also why the object has the color that it does. If it is a red brick, the red is reflected from the incoming white light (with some other surrounding wavelengths) while the others are absorbed.

Returning to Light passing though a material and its shift in direction is this idea, the **Law of Refraction** is also known as **Snell's Law** (aka **Descartes' Law**, the **Snell -Descartes Law**) *states that the ratio of the sines of the angles of incidence and of refraction is a constant that depends on the media*. The constant turns out to be what is called the **Index of Refraction** for a given media. It is used as a characteristic for a material and is used in science regularly.

It was discovered and rediscovered hence the reason for different names and people from different times. One of the main people is Dutch mathematician Willebrord Snellius in 1621. He writes his law in this manner : the ratio of the sines of the angles of incidence and refraction is equivalent to the ratio of the velocities in the two media (or equivalent to the opposite ratio of the indices of refraction of the two media).

$$\frac{\sin\theta_i}{\sin\theta_r} = \frac{v_i}{v_r} = \frac{\lambda_i}{\lambda_r} = \frac{n_r}{n_i}$$

Θ is angle, i = incident, r = refracted, v = speed, λ = wavelength, n = index of refraction

Snell's Law basically follows from **Fermat's Principle of Least Time**, (the principle states that the path taken between two points by a ray of light is the path that can be traversed in the least time. It is often used as the definition of a ray of light – a more modern view is that rays of light traverse the path of stationary, not minimal, time) The principle can be used to describe the properties of light rays reflected off surfaces or refracted through different media, and even those undergoing total internal reflection. Hence this principle treats light as a wave in its propagation. The law is used to determine the direction of light rays through refractive media of varying indices of refraction.

The next thing to consider is this : since the wave speed changes, how does this affect the way the wave bends? Let's make the waves into a different model where each wave is a set of wheels on an axle. In a given medium, they all travel at the same speed. Once passing into a new medium, much like having a tile floor for the wheels to roll across and now the wheels encounter carpeting, the speed changes in the new medium.

Using the wheel and axle model described here, if going from tile (the faster speed medium) to the carpeting (the slower speed medium) the wheel that contacts the carpeting first slows a bit, while the other on the tile rotates at a faster rate, hence the axle system changed direction and for the waves, it results in a wave that bends towards the normal. If going from the slower medium (carpeting here) to the faster one (tile here) the wheels pick up speed and bend away from the medium.

A quick aside is needed here. Many have read of the idea that the speed of light is absolute and this is one of the cornerstones of Einstein's Relativity Laws. Yet here there seems to be a contradiction?! No, this is not the case. The speed of light actually remains intact as the wave passes through various media. What happens, for example, when a light wave encounters plane glass is that the wave is absorbed by electrons in the atoms, which in turn become excited and these spontaneously emit these photons which are absorbed in turn by the next set of atoms in the glass and so on when these waves collide with the atoms they encounter. When the wave travels from one atom to the next, it is traveling at the speed of light. But from the measurements of the distance it travels (the thickness of the glass) and the time it takes to enter and exit, it would seem that the wave had slowed down. Funny thing, though, instantly upon exiting and emitting from the opposite surface atoms it is again moving at the speed of light! What has happened is a time delay in getting through the material. It has to be absorbed and emitted. There is a time delay in this process. Speed is distance divided by time. The distance is known and remains as such, but the time takes longer than it would if there were no barrier at all. Hence there is no violation and the speed of light remains as constant as ever.

There is a value associated with a given medium to describe this affect. It is called the Index of Refraction (n) for a medium. A vacuum has a value of n=1. For most calculations in a classroom, air is taken to be n=1 also (though it is n=1.0003 or something close to that for purists). Regular water has a value of n=1.33. The value of 'n' is affected by the density of the medium.

Can the wave start in a medium of higher index of refraction and head into one of a lower index? Yes. In fact, all ray paths (as long as the conditions are identical) are retraceable. This is why objects under the water appear closer to the surface than they actually are! The light path from the rays of light bouncing off of the object are refracted when passing from water into the air.

Another interesting thing can occur too. What if the light rays going from the medium of higher index of refraction into the one of a lower index of refraction exceed 1 in the ratio of the indices? Why is 1 so important? The sine of any angle cannot exceed 1, that is why. Does the rule break down for these cases? No. In fact, what happens is that the light ray is *totally internally reflected*. This angle has a special name and is called the **Critical Angle**. Instead of refraction taking place, the beam remains within the medium entirely. This property of a medium is taken advantage of in the case of fiber optics. Light rays propagate in a straight line, yet fiber optic cables with a beam of light going through them can readily be bent. This is because of their index of refraction exceeding the critical angle as it is called so that all of the light remains inside the fiber optic cable.

A quick question from the observant might arise : does the index of refraction depend on the wavelength of the wave? Yes it does. The shorter the wavelength, the greater the index of refraction from measurements. For example when rating materials for their indices of refraction (actually done for glass and used in various sciences, Physics, Chemistry and Forensic Science) there will be a number, for example n=1.5 for a given glass make. Next to it the statement will note for which wavelength of light, 500 nm. In this case, if red light (695 nm) were used the index of refraction might be less, like 1.49, while at a higher frequency (shorter wavelength) it might be 1.51 for blue light (405 nm).

There is a name for the phenomena of white light being separated into its constituent wavelengths (hence colors) due to this dependence of the index of refraction on wavelength and it is called **Dispersion**. This is the phenomena studied and noted by none other than **Sir Isaac Newton** and began the science of optics. It is also in part the reason for rainbows. Yes the raindrops disperse the light, but this is after the waves undergo internal reflection off the drops.

Lenses & Mirrors

Though Optics is but a branch of Electromagnetic Waves, which in turn is a subset of Waves and this is but a branch of Physics seems therefore like a small topic, it turns out to have the most panoramic and all-inclusive history in science as we shall see. Lenses along with Mirrors have been at the foundations of various branches of science and have paved the way for innumerable discoveries.

Though lenses have been around since the 1st century (about the year 100) where glass existed. Some with thicker middles and thinner edges were shown to magnify an object placed under it and looked at through it. This lead to what we still call magnifiers which could be used to focus sunlight to start a fire.

The first lenses did not extend beyond this point for many centuries until the 13th century where spectacle makers became more and more prevalent and glasses to be worn were crafted.

The first serious work with lenses occurred in 1597 when Zaccharias Janssen and his son Hans, two Dutch eyeglass makers, experimented with lenses in tubes. They inadvertently constructed not only the first microscope, but the first compound one (having more than one lens). It's power was rather limited (10x).

Reports of this use of lenses and some by others made their way to Galileo who went about the task of crafting his own lens systems. Galileo did make a compound microscope, but it was only useful for reflected light. Robert Hooke, reading on these things some time later, created the first English compound microscope in 1655.

Not only did Galileo use lenses for the microscope, but he crafted lenses for his spyglass, the refracting telescope. This idea came from not only Zacharias's work, but also Hans Lippershey in 1608 who is credited with the first working telescope. Like Lippershey, Galileo's scope used a convex objective lens and a concave eyepiece lens. This item brought the universe to us and changed history forever. Galileo saw the phases of Venus, mountains on the Moon, the moons of Jupiter, and an unusual shape to Saturn (later found to be its ring, of course). Also he found that the Sun had spots. This along with his publication of the discussion of the Heliocentric Theory in his book, the Dialogue, opened up the realm of science and pitted him as adversary to the powers of those times. This was the beginning of Astronomy in every measureable way and continues to this day. The importance of the optical telescope cannot be underestimated. Consider the discovery of Uranus, Neptune, and Pluto, not to mention all of the work by later astronomers such as Hubble who uncovered that island universes were galaxies, large masses of stars, at fantastic distances and more importantly, moving away from us. After him, there are the countless discoveries of the Hubble telescope names after him.

Lenses did not stop there, however. Dutch lens grinder Antoine van Leeuwenhoek created single lens microscopes in the 1670s. His process made millimeter size lenses that could magnify on the order of a few hundred times. Attached to a small brass instrument and guided by screws and pins. His work led him to find and label the cell. This accompanies Hooke's own discoveries as reported in Micrographia (1665). Hooke confirmed van Leeuwenhoek's work and examined cork pieces noting the tiny chambers and coined the term cell. The cell theory today is the foundation of biology. The 1700s and 1800s and even into the 1900s, the compound microscope was the biologists main lab tool to look at still smaller and smaller scale at the structure of the parts of all living things. These explorers were looking at the realm of the small and finding life and structures everywhere.

Lenses continued to progress and change through time, though. Kepler first improved the Galileo telescope design by using two convex lenses. By 1655 Christiaan Huygens were constructing ever-more powerful telescopes. It was also Johannes Kepler, the mathematician and astronomer who first determined more accurately a way to analyze eye problems and craft even better glasses. Years later, this was followed by Ben Franklin who took it a step further and developed what we still call today as bifocal lenses, where one lens corrects for a vision problem while the other allows use for distant viewing regularly.

The Lens is not the end of the story, though. The Mirror has its own history. Though mostly used as a reflecting material and when warped from a flat plane can be amusing in a funhouse, if mathematically done and employed in certain endeavors, it leads to powerful tools.

Even the most renowned scientist of all time, Isaac Newton, studied light. His first major paper, On Color, was about the refraction of white light into its component and independent colors of the spectrum (ROYGBIV). This paper was rejected and ridiculed by Hooke, hence Newton became very withdrawn and did not publish his greatest work, the Principia for some time (1687).

Also, it was none other than Isaac Newton who had read about Galileo's telescope, but noted the shortcomings of light passing through glass (chromatic aberration, et al) and wanted a better means to examine starlight. He devised a concave mirror to capture as much light as possible and its curved surface would focus the light to a point to be analyzed through viewing. Hence was born the Reflecting Telescope in 1668 (like the ones found in most Observatories today and is the type that the Hubble telescope is). We even call this type the Newtonian Telescope today. Reflecting Mirror Telescopes can be much larger than Refracting Lens Telescopes and hence show much more detail in the sky. For Newton, this was part of the reason for his acceptance into the Royal Society.

In fact it is Newton who publishes later in his book Optics ideas on the nature of light, experiments conducted and those that need to be explored, a reprinting of his early paper on the spectrum of light, plus in it a summary of his discovery of calculus (fluxions). This lead to the debates of the wave and particle nature of light, which was won for some time by Huygens where he demonstrated wave properties from diffraction (see Wavelength of Light Activity #6).

The idea of light was later revisited by Einstein who views light as both a particle and a wave and it depends on the experiment on how its nature is measured. Light is found to be the constant of the universe, that is the speed of light is its limiting speed and only energy can travel at this and matter can approach it but cannot attain it. The speed of light was analyzed by utilizing properties of optical instruments such as mirrors and prisms to separate and reconstruct a beam of distant light to try and detect any variation in its speed through space (which it did not hence showing there is no medium in space and electromagnetic waves do not need one) [this is called the Michelson-Morely Experiment]. Note : Earlier experiments by Albert Abraham Michelson as well as Simon Newcomb determined a more accurate value for the speed of light.

Lenses and Mirrors have been at the center of the foundations of biology, the advances in astronomy, and has expanded our horizon by large orders of magnitude.

The main equation used for analyzing both mirrors and lenses is what is classically called the **Thin Lens Equation** (Formula) (aka the Lens-makers or glass-makers formula). It relates the distances of both the object and the image to the focal length of a given mirror or lens. The key to using it is understanding whether the focal length is negative or not which then creates an expectation of where the image forms and what type of image it is.

$$\frac{1}{f} = \frac{1}{o} + \frac{1}{i}$$

i = image distance, o = object distance, f = focal length

Note that the lens equation formula applies to 'thin lenses'. What this means is that the thickness of the lens with respect to the focal length and the distances involved is fairly small. The thinner the lens with respect to these values, the better the estimation of the equation to the actual outcome. There are other formulae for thicker lenses and more complicated situations, but these are not explored here.

Despite their differences and complexity, whether it is a Mirror or Lens it will form one of two types of images : Real or Virtual. (While reading this section through make the connections between ideas and re-read it so as to fully understand it). A **Real Image** is one that is formed by the intersection of light rays and can be projected onto a screen. A **Virtual Image** is one formed in which light rays appear to diverge from, even though they are not actually focused there and this image cannot be projected onto a screen.

Two other classifications apply to images : **Upright** or **Inverted**. Upright Images are situated in the same manner as the original object while Inverted Images are just that, upside down as compared to the original Object.

One other possibility remains for an image : **Magnified** or not. Magnification is not only an increase in size (where M > 1), but it can be a decrease in size in the image as compared to the object (0 < M < 1). From the formula it can be seen that the Magnification can be negative?! The sign is actually a tool to help determine whether the object is upright and virtual or inverted and real.

$$M = -\frac{i}{o} = \frac{h_i}{h_o}$$

Sign of M	Type of Image	Image compared to Object in Orientation
+	Virtual	Upright
-	Real	Inverted

The Law of Reflection was discussed in the Law of Refraction Activity (for comparison/contrast purposes) and also here there is little note of a Plane Mirror, the most commonly used form. In the case of a **Plane Mirror**, here are the basic facts : The image is always the same distance from the mirror as the object and is of the same size (hence no magnification). If one can see another in a reflection in the mirror then that person too can be seen. That is to say, ray paths are retraceable. The reason that an image is left-right reversed and not top-bottom reverse (your image raises its left hand when you raise your right hand) is because the image is inverted front to back and NOT side to side. The person is facing you in the mirror and not away from you as you are standing in relation to the mirror!

The two main types of (spherical) Mirrors are : the Convex Mirror and the Concave Mirror.

The Convex Mirror :
A Concave (spherical) Mirror is one who reflecting surface is an outward-curved segment of a sphere. This is like the back of a polished metal spoon. Since the Object is in front of the mirror, its distance is positive. The Focal Point is behind the mirror, so it always negative. This type of mirror always creates a virtual image between the mirror plane and the focal point, hence the image distance is negative. The image is upright and always smaller than the object (M<1). Reflect on the formula and you can see why these results come out. Since 'o' is always +positive and 'f' is always −negative, the answer for 'i', the image distance will be −negative and 0<i<f. To illustrate, return to the spoon example noted above and move your finger towards and away from the back of a polished metal spoon to find that the image is always upright and smaller.

Convex Mirrors have some uses. Small ones are attached to plane mirrors on trucks, buses, and large vehicles. This is used to provide a wider field of view, though it is small in dimension. Larger curved mirrors like this are used in stores at the intersection of aisles and hallways since it allows a field of view in a number of directions.

The Concave Mirror :

The Concave (spherical) Mirror is one whose reflecting surface is a segment of the inside of a sphere. That is to say it curves inward. This is like the 'bowl or cup' portion of a polished metal spoon. Unlike the concave mirror, the image produced here will vary with placement of the object. The focal point distance is in front of the mirror and is always positive. Also, the object distance is positive (if it is in front of the mirror).

A Concave Mirror can form Real or Virtual Images. Also these images may be upright or inverted, also smaller or larger than the object itself. All of this, again, depends on the object's initial location with regards to the radius of curvature of the mirror (2x the focal distance), the focal distance, and the mirror itself.

This paragraph will have a list of summaries for the images formed by a Concave Mirror as the Object is placed first far away and then brought towards the mirror.

1) If the Object is at 'infinity', it will form an incredibly small real image all at the focus of the mirror itself. (This idea is employed by telescopes).
2) If the Object is at a reasonable distance, yet further than the radius of curvature, it will always form an inverted, smaller (M<1) real image (real images here will be in front of the mirror) between the focal length and the radius of curvature.
3) If the Object is at the radius of curvature distance, it will form a real, inverted image at the same distance of the radius of curvature where it is the same size as the object (M = 1).
4) If the Object is between the radius of curvature and the focal point distance, it will form an inverted, real image at greater than the radius of curvature distance with increased size (M > 1).
5) If the Object is at the focal point distance, its image is projected into infinity and there is no image formed.
6) If the Object is between the focal point distance and the mirror itself, it will form an image that is upright, virtual (behind the mirror), that is larger than the mirror itself (M > 1).

To test this idea, move your finger towards and away from the cup of a polished metal spoon and find that it changes size and orientation a great deal.

These concave mirrors are used a great deal as they are the basis of all modern telescopes (see above in history with Isaac Newton).
Much like Mirrors, there are two basic Lenses to consider : the Diverging (Concave) Lens and the Converging (Convex) Lens. First, a Lens is a transparent object that refracts light rays such that they converge or diverge to create an image.

The Diverging (Concave) Lens :
A Diverging (Concave) Lens is thinner in the middle and thicker on the edges.
It always produces a virtual image of any real object regardless of placement.
The image is upright and the size is always reduced (M < 1). The image
always appears inside the focal point as well. This is somewhat parallel to a
convex mirror. For a diverging lens the focal length, f is negative.
The primary use of this lens is in the correction of nearsightedness.

The Converging (Convex) Lens :
A Converging (Convex) Lens is thicker in the middle and thinner on the edges.
The best example is the handheld magnifying lens.
This paragraph will have a list of summaries for the images formed by a Convex
(Converging) Lens as the Object is placed first far away and then brought
towards the lens.

1) If the Object is at 'infinity', it will form an incredibly small real image all at the focus
   of the lens itself. (This idea is how sunlight is focused and burns a hole in a piece of
   paper with a magnifying lens).
2) If the Object is at a reasonable distance, yet further than twice the focal length, it
   will always form an inverted, smaller (M<1) real image (real images here will be in
   back of the lens) between the focal length and twice the focal length. (the
   application of this is the lens of a camera, the human eyeball lens, and the objective
   lens of a refracting telescope).
3) If the Object is at twice the focal length distance, it will form a real, inverted image
   at the same distance of twice the focal length where it is the same size as the object
   (M = 1). (Note, : being a real image it is on the other side of the lens).
4) If the Object is between twice the focal length and the focal point distance, it will
   form an inverted, real image at greater than twice the focal length distance with
   increased size (M > 1). (the application of this is in slide projector objective lenses
   and compound microscopes).
5) If the Object is at the focal point distance, its image is projected into infinity and
   there is no image formed. (the application of this is in lighthouses and search lights)
6) If the Object is between the focal point distance and the lens itself, it will form an
   image that is upright, virtual (in front of the lens), that is larger than the mirror
   itself (M > 1). (Note : This is how we use this lens as a magnifier). (the
   application of this is in the eyepiece lenses of binoculars, microscopes, and
   telescopes).

What is important to note in all of these discussions is this : converging lenses
can be compared to concave mirrors and diverging lenses can be compared to
convex mirrors where the sign conventions and images formed are analogs.

Section on Converging Lenses

Lenses come in a variety of sizes but the most recognized one is what is called the double convex or converging lens. A simple definition for a converging lens is one that is thicker in the middle and thinner on the edges. Like any other lens this can be made of glass, plastic, or other transparent materials. This is the one classically used as a magnifying lens and stereotypically in the hands of fictional detectives such as Sherlock Holmes along with scrutinizing scientists.

Oddly enough, the use of a converging lens as a magnifier requires that the object under study to be inside the focal length of the lens itself! An object that is in front of the lens and inside the focal length distance will be produce an image that is in the same direction as the original object and larger than the original object. Because the Image is on the same side of the lens as the Object means that it is a Virtual Image.

In the case of the converging lens its convex sides are made so as to take advantage of the behavior of rays passing through transparent materials called refraction and due to the surfaces geometry, the rays are bent in order to converge. The greatest curvature is at the edges, hence the greatest amount of bending of the wave.

Distant parallel wave fronts approaching this type of lens (that also means parallel rays to the principal axis of the lens) are focused to a point called the Focal Point. The distance from the center of the lens to the focal point is the Focal Length of the lens. The focal length is positive for sign conventions in calculations.

These lenses are common in nature as these are the type in our eyes and in other life forms with complex eye systems. In our eyes, the retina is at a fixed distance from the lens. The lens itself is not a fixed item. It has flexibility that allows the lens to have an adjustable focal length through contraction and relaxing of the ciliary muscles. This flexibility typically changes with time in the lens itself.

The two most common optical defects are nearsightedness and farsightedness. These were noted by **Johannes Kepler** who helped create some of the first glasses in history (see Ray Diagram Activity for more history).

In nearsighted (myopic) eyes, the focal length is too short hence the distant object comes into focus in front of the retina and does not form a sharp image on the retina. Here glasses with diverging lenses are used to correct this condition.

In farsighted (hyperopic) eyes, the focal length is too long, hence the nearby objects converge at a point beyond the retina and so will not form a sharp image on the retina. Here, eyeglasses with converging lenses can help correct this.

The converging of lens has a number of uses. A camera lens is of this type as it focuses the image onto the film. Not only this, but this type is used for films and slides (old style and even modern power point) through projectors. These are Real Images on the screen. There are many other devices, such as binoculars (also including other lenses and even prisms to bounce around the image to bring it to the eyes for viewing).

Optical Microscopes also use this lens as the primary tool for imaging an object. The compound microscope even employs two converging lenses to image an object and produce a magnified real image.

Optical Telescopes, like the one **Galileo** used, employ this type of lens. Since not all rays converge at the same point (called **chromatic aberration**) other lenses are put in with this main lens to create a clear and crisp image. The largest one is the **Yerkes telescope** which is 40" in diameter. Though small as compared to mirror telescopes (up to and more than 200+") this is due to the fact that unlike a mirror which can be supported from the back and has only one surface to polish to a high degree of smoothness, a lens is supported on its edges and must be polished on both sides.

Waves Conclusion

Waves are indeed a critical part of our universe, our world, and our lives both in the realm of phenomena and in the application of uses. We use the radio portion of the electromagnetic spectrum for communication and data transmissions. Many of the other waves in the electromagnetic spectrum are used for an array of uses, x-rays and microwaves are quick examples. Waves are used for analysis in many sciences from biological and medical to geophysical and astronomical. Waves are transporters of energy. With efforts at using them, we will find ourselves tapping into them for energy sources in the present and future. All the best in your explorations and studies of waves through these Activities and your reading. Enjoy. :)

Activity #1
Using Math (and the Slide Rule) in Everyday Life
Grade Level : Middle School
Math Level : Calculating

Everyday Life Calculations with the Slide Rule Activity

This Activity is the mathematical exploration of everyday life – such as determining miles per gallon, cost per unit ounce and the like, but all using a slide rule. Even if you don't, though I recommend it, these are basic things we should all know and engage in mentally to some degree. It is good to practice common math sense! :)

- All of these calculations require either a question on the part of the person to consider the hypothetical or the actual items at hand to work on for calculations.
- The materials needed depend on what is being done.
- Know the basic rules for multiplication and division on a slide rule (they are summarized below and if needed in the first question – miles per gallon).
- When scales other than C & D are used the rules are explained.
- In most cases, only the C scale and D scales are needed.
- Always estimate an answer and watch decimal placement.
- Not always needed but is handy is a small pad of paper.

Slide Rule Basics

1) For almost all calculations in this Activity, the C & D scales are used.
2) If considering a Ratio or Fraction, It is best to see the C Scale as the Numerator since it sits atop the D Scale and the D Scale as the Denominator
3) To Multiply, Place the Right or Left Index as needed of a given scale, say the D Scale below the first number to be multiplied on the C Scale.
4) Next, read along the D scale to the other number to be multiplied by and read above it the answer on the C scale!
5) If the number is not available, use the other index (Right or Left) and start the process over as needed.
6) If Dividing simply place the Numerator value on the C Scale over the Denominator Value on the D scale.
7) Read the Answer opposite the Index of the D Scale on the C Scale.
8) What about the use of Scientific Notation?
9) It is best to consider all values in Scientific Notation in all calculations. This is done by setting the decimal point past the first non-zero number in the value under consideration and multiplying by 10 raised to the appropriate power.
10) The power is determined to be positive if the decimal needs to move to the right to obtain the original number (2000 is 2×10^3 for example), and negative if the decimal needs to move to the left to achieve the original number (0.002 is 2×10^{-3} for example).
11) When multiplying with values in Scientific Notation, simply add the exponents (Result = Operand 1 Exponent + Operand 2 Exponent).

12)

13) When dividing by Values in Scientific Notation, simply subtract the Denominator Exponent from the Numerator Exponent (Result = Numerator Exponent – Denominator Exponent).

14) Note the special case rules for linear slide rules :

15) If Dividing and the Slide Projects Right, Then Subtract One from Your Exponent Total to achieve the correct decimal placement.

16) If Multiplying and the Slide Projects Left, Then Add One to Your Exponent total to achieve the correct decimal placement.

17) Here is a full summary of decimal placement :

> **The Rules for Decimal Placement :**
> 1) Always first estimate the Answer.
> 2) Convert all values into Scientific Notation.
> 3) For Any Multiplied Values, Add up the Exponents and For Any Divided Values, Subtract the Exponents.
> 4) If for any single Multiplication operation the Slide moves to the Left, then add +1 to the exponent total
> 5) If for any single Division operation and the Slide moves to the Right, then -1 from the exponent total
> 6) Mentally treat any two values independently (just as whole numbers now) but keep track of what the exponent will be through the rules.
> 7) That is to say answer the question as to where the Slide has moved and factor in that addition or subtraction as needed for each operation independently.
> 8) Take this final answer and use the exponent figure arrived at to determine the decimal placement!

Miles per Gallon Measured

1) For this calculation it is best to begin with a full tank of gas. Once full write down the number of miles on the Odometer [called the starting miles] (or reset the trip odometer to zero).

2) Drive for some period of time (it can be a day, but if you have a regular work schedule it may be better to go for 2 or 3 days).

3) Needless to say you put more gas in some time later. At that time write down 2 numbers.

4) First the current reading of the Odometer now [called the ending miles] (or look at the trip odometer and write down the number of miles driven).

5) Second write down the exact amount of gas put in this time. This is the number of gallons used.

6) The 10" slide rule will give reasonable results despite the number of digits in your numbers, but it is best to round off as needed to facilitate the calculation.

7) If you did not use the trip odometer, determine the difference between your starting miles and your ending miles by subtracting them. This is your trip total.

8) Quick review for Division and Multiplication on a Slide Rule :

9) To Divide place the numerator read from the C scale over the denominator on the D scale and read the answer opposite the D scale index on the C scale.

10)

11) To Multiply place the left (right if needed) Index of the C scale over the first operand read from the D scale. Read along the C scale to the other operand and read opposite it on the D scale to find the Answer.

12) To determine the <u>average miles per gallon</u> :

$$\mathbf{Mpg} = \frac{\textbf{trip total}}{\textbf{number of gallons used}}$$

13) This value is important in many ways – On long trips and the car is gassed up, you can now estimate the range the car will go before filling up by *multiplying the Mpg by the number of gallons.*

14) On a map, knowing the distance to some place an estimate of the amount of gas needed can be determined by *dividing the Trip Miles by the Mpg.*

15) Also estimated costs can be found by taking *the number of gallons used and multiplying by the average cost of gas at that time.*

16) This calculation can go along with the Average Speed one as well.

Average Speed along with Distance and Time Determinations

1) To find the average speed what is needed is the total distance travelled and the total time taken for the trip.

2) This calculation can be done for cars, bikes, walking, boats, or any moving object under consideration.

3) If by car, either determine the total distance to be traveled by examining a map (paper for those who like to figure it out themselves or on the internet and a map system like Google Maps) OR

4) Actually travel the distance and either reset the trip odometer before the trip or write down the current odometer reading (starting miles) and then at the end of the trip write down the new odometer reading   (ending miles).

5) In either case of the trip, it must be traveled and times.

6) For the traveled trip be sure to not only keep track of miles but use a watch or a timer. On a watch mark down the start and end times.

7) Determine the miles traveled by subtracting the starting miles from the ending miles (Miles Traveled) and

8) Do the same for the amount of time to travel by subtracting the start time from the end time (Travel Time).

9) The Average Speed is found by :

$$\mathbf{v} = \frac{\textbf{Miles Traveled}}{\textbf{Travel Time}}$$

10) Note that one does not need the total miles and total times to determine an average speed for part of a trip.

11) If you are on a long trip and had the start time and start miles down, then at any point in the trip the average speed for that portion of the trip can be determined.

12)

13) This might be useful in the case of a very long drive where if the total miles to cover is known and the average speed is determined for some point in the trip, then the amount of time needed is found by :

14) First subtract the miles covered from the total miles. *Take the remainder and divide this value by the average speed. This will give the amount of time in hours left for the journey.*

15) Estimated times can be determined for an entire trip by first estimating the average speed and dividing it into the total miles of the trip. This gives the number of estimated hours for the trip.

16) If there is a situation where the average speed is known and the amount of time is also known then *the distance traveled is simply the average speed times the amount of time traveled.*

17) Always watch to see that units match up as needed! If not convert units to agree with the problem at hand.

18) A good exercise is to convert units for those who like math practice. For example, convert mph to miles per second or feet per second.

19)   Miles per Second = mph / 3600

20)   Feet per Second = Miles per Second * 5280

21) Also why not try the metric conversions as well, such as mph to km/hr and then convert these to m/s.

22)   Kilometers per Hour (km/hr)= 1.6(09) * Miles per Hour

23)   Meters per Second = (Km/hr) / 3.6

24) These ideas apply to taking a flight too. Know the distance flown and time the trip to determine the average speed of the plane.

25) Note that distances do not have to be miles, it can be any measurement unit. A bicycle can be looked at from the number of houses it passes in a given time, for example. A crawling insect might be done in centimeters/minute.

Cost per Unit Known (Mass, Volume, Weight, Item Number)

1. Most stores today will provide the cost of an item per unit ounce, per pound, per unit number of the item. The first question then to determine is this : Is it correct?
2. The next goal of this activity is to compare it to other brands or the same brand on the shelf only packaged in a different size.
3. Note that often the larger sizes are not labeled in the same manner of cost per unit as the smaller ones.
4. Be sure to read the labels carefully since the unit cost provided might be something like cents per ounce yet the material is labeled in dollars and pounds.
5. Therefore it is a good idea to keep in the back of one's mind basic conversions, such as :
   - **1 pound = 16 ounces**
   - **1 cup = 8 ounces**
   - **1 gallon = 4 quarts**
   - **1 quart = 2 pints**
   - **1 pint = 16 ounces**
   - **1 kilogram = 1000 grams**
   - **1 kilogram = 2.2 lbs (pounds)**
   - **1 ounce = 28.3 grams**

6. With these in mind it is best to choose base units for calculation and comparison, for example turn all units to ounces for mass or volume considerations.
7. Reading a label first determine the total number of ounces if not given.
8. To calculate the cost per unit ounce (CPO):

9. **Cost per Unit Ounce (CPO)** $= \dfrac{\text{Cost (\$)}}{\text{No. of Ounces}}$

10. For all other types of costs per unit whatever it is simply the cost divided by the number of ounces (number of pounds) (number of items) and the like.
11. In all cases these are rates and a rate is a ratio or fraction.
12. The best way to treat all ratios is to let the Numerator (here Cost) be the C scale and the Numerator be the D scale (here the number of Ounces). The answer is read opposite the D scale index on the C scale.
13. Once you know the cost per item it is easy to determine the total cost of a purchase by taking that ratio and multiplying by the number of items you intend to buy.

Computing Sales Tax or Determining the Tip

1) This activity is not a mistake though these two things are different circumstances. It turns out they are calculated in the same way on the slide rule.
2) For both you are mathematically taking a cost and adding to it a percentage of the cost.
3) Read the Cost on the C scale and place the left index of the D scale beneath it.
4) Now read along the D scale to the known sales tax rate or the desired amount of tip.

5)
6) Here each mark on the slide rule represents 1% or 1/100.
7) For example, The second mark is 2%, the fifth mark is 5%.
8) The easy mistake is misreading the D scale. Be sure to keep track of the decimal point here.
9) For example if the cost were $20 the first mark past the left index on the D scale has a reading of $20.20, the fifth mark is $21.00
10) Yet if the cost had been $200, then the first mark has a reading of $202 and the fifth mark has a reading of $210.
11) This illustrates the power of the slide rule since all numbers can be represented on this logarithmically-spaced number line and it is merely the decimal point that needs to be tracked.
12) When you read to 1.1 this is 10%. The value above it on the C scale is 10% greater than the original value.
13) If you want the amount of tax or tip you have to subtract the original cost from this new value to see how much the tip is.

Calculating Cost of a Sale Item (given percent off)

1) Many of us have been to stores proclaiming some given percentage off and large tables on cards describing what one pays.
2) With a slide rule this is a very easy task.
3) First, we need to acknowledge the trick in advertising. If it is 10% off, we are still paying 90% (100%-10%), and we pay 75% if it is 25% off (100%-25%) and so on.
4) Knowing this makes the task of finding our actual cost very easy and we can even find our savings too.
5) Read the Original Price on the C scale and place it over the Right Index of the D scale.
6) Now read from right to left along the C scale and find the percentage you are paying.
7) For example, if you have 10% off, read to 9 which is now considered 90% and read above it on the C scale your price.
8) If you have 25% off read to 7.5 since this is 75% and read the price above on the C scale.
9) What if you want the amount of savings? Instead of the right index, place the Price over the Left Index of the C scale.
10) Reading from left to right find the percentage off and find its value on the D scale above.
11) Note that you may have to use the Right or Left Index differently if the values go off the scale!
12) Notice the speed of this reading. In a single moment one can determine what any percentage is for any given value. Try that one with a calculator for those who think that faster!
13) By the way go back to the sales tax activity above if there is tax on it to determine the total cost if needed.

Computing Pay Amount from Rates of Pay

1) For estimated gross pay, take the rate of pay (R) on the C scale
2) Place this value over the index on the D scale
3) Read along the D to find the Number of Hours worked (H)
4) Find the answer on the C scale above (G)
5) Notes : First note that the pay rate has to be in the same time unit as is multiplied. Second, this is the gross without any withholding. To factor that in take an estimated percentage of pay withheld for taxes and use the percent off calculation to find an estimated Net Pay. Third, if there is overtime pay, be sure to calculate those hours at that pay rate and add that to the base gross pay.

G = R*H

Estimating the Electric (Water) Bill from Meter Readings

1) This activity here examines directly the process of reading the meter, taking the values there and finding an estimated bill (excluding taxes and other fees on the bill).
2) Depending on the meter you have for the Electric and Water Meters, this will affect how it is read.
3) There are some that give a direct reading of the current level of consumption, and if present then you can proceed with the calculation.
4) If the number is not directly on it, then you need to start on the first day of the month if you want to monitor it for the month. (If for a week, pick any day).
5) Read the dial carefully. Most have dials and these go in opposite directions with each dial, first clockwise, then counterclockwise, etc.
6) When the arrow is between numbers be clear on how to determine what value it is. The arrow is always the value it is coming from and until it reaches the next value on the dial.
7) If the meter is in a dial pattern, be sure to start with the largest place value and work your way around the dial for the other digits.
8) In both the Electric and Water Meter cases, you need the starting value and then an ending value some time later (keep track of the time between your readings – a month is a good time frame).
9) Simply take the Final Number you record and subtract the First Number you record. This Net Amount Used is what you consumed.
10) In the case of Electricity it is normally in kilowatt*hours (kWh) while in the case of Water it is 100s of cubic feet of water (Ccf).
11) Strangely, the Water Company typically charges for Ccf and not gallons. For those who want to know the gallons, take the Ccf number and multiply by 748 as there are 7.48 gallons in a cubic foot)
12) In both cases, one can call or look up on line the average cost per kWh or Ccf and then simply multiply your use by these values respectively to arrive at an estimated cost.

Cost = Net Amount Used x Rate

13) Note that this cost is estimated, since it does not factor in any type of tier-pay system for cost changes for different amounts, any taxes, fees, and other costs that may appear on the bill.

Home Project Needs for Painting and its Cost

1) The most basic of calculation is simple enough, first measure the Length (L) and Height (H) of a wall in a room in the chosen units (ft or m) and calculate the Area (A). (**A = L * H**)
2) To make the process simpler and avoid mistakes round all values up to the next whole number value. Most walls are 8 ft tall for example.
3) Be sure to sum up all of the walls. TA = total area (**TA = Sum of all Areas**)
4) Not Recommended , But For those who like some level of precision, in the case of inches divide the inches past the feet measure by 12 and tack this decimal value onto your feet measure, instead of the rounded up values used.
5) For still greater precision, you can subtract out non-painted areas, such as windows and doors (A = L*W) if you like.
6) Whatever the case, take the final number TA and divide by 300. This gives the number of gallons needed for rough, textured, or unpainted wallboard.
7) If it is smooth walls instead divide TA by 350.
8) Note that these are estimates. Always overestimate and round up. For example you are not going to buy 3.3 gallons, buy 4.
9) The final calculation is cost, simply take the Cost per gallon on C scale place over 1 on the D scale and read along the D scale to the needed number of gallons. The Cost is found on the C scale above.

Home Project Needs for Wallpapering and its Cost

1) The most basic of calculation is simple enough, first measure the Length (L) and Height (H) of a wall in a room in the chosen units (ft or m) and calculate the Area (A). (A = L * H)
2) To make the process simpler and avoid mistakes round all values up to the next whole number value. Most walls are 8 ft tall for example.
3) Be sure to sum up all of the walls. TA = total area (TA = Sum of all Areas) This is the Wallpapering Area.
4) For still greater precision, you can subtract out non-papered areas, such as windows and doors (A = L*W) if you like, instead of rounding up. Subtract the total of the non-papered areas from the papered ones
5) Use the chart below to find the Usable Yield Value. This is divided into the Wallpapering Area to determine the Number of Single Rolls Needed.

$$\textbf{No. of Single Rolls} = \frac{\textbf{Wallpapering Area}}{\textbf{Usable Yield}}$$

6) The final calculation is cost, simply take the Cost per roll on C scale place over 1 on the D scale and read along the D scale to the needed number of rolls. The Cost is found on the C scale above.

Pattern Repeat (Drop)	Usable Yield (American rolls)	Usable Yield (European rolls)
0 to 6 in.	32 sq ft.	25 sq ft.
7 to 12 in.	30 sq ft.	22 sq ft.
13 to 18 in.	27 sq ft.	20 sq ft.
19 to 23 in.	25 sq ft.	18 sq ft.

Home Project Needs for Carpeting and its Cost

1) The most basic of calculation is simple enough, first measure the Length (L) and Width (W) of the room in the chosen units (ft or m) and calculate the Area (A).
($A = L * W$)
2) To make the process simpler and avoid mistakes round all values up to the next whole number value.
3) *Not Recommended But For those who like some level of precision, in the case of inches divide the inches past the feet measure by 12 and tack this decimal value onto your feet measure, instead of rounding up.*
4) Determine the Number of Square Yards Needed by taking the Total Area (in square feet so far) and divide by 9.

Number of Yards Needed $= \dfrac{\textbf{Total Area in sq ft}}{\textbf{9}}$

5) Note that this Calculation is for Vinyl Flooring too! Follow the same procedure noted above down to the cost here below.
6) The final calculation is cost, simply take the Cost per yard on C scale place over 1 on the D scale and read along the D scale to the needed number of yards needed. The Cost is found on the C scale above.

Determining Recipe Needs through Proportions

1) These directions addresses the question of what to do when a recipe does not match your materials on hand and/or the question of how much is needed when you are changing the scale of the recipe to a larger or smaller yield.
2) This is the type of activity that the slide rule does very well since it is a proportion, which slide rules excel at.
3) The idea to solve this comes from noting that the ratio of the needed amount for a given ingredient in the recipe of a given number of servings will equal
4) The amount you need to have for your batch divided by the number of servings you wish to make!

$$\frac{\textbf{Recipe Requirement for Material}}{\textbf{Recipe Number of Servings}} = \frac{\textbf{The amount of Material Needed}}{\textbf{Number of Servings}}$$

5) This type of logic can be used for any and all variations of the same theme here. Remember to let the Numerator be the C Scale while the Denominator is the D Scale.

6) First place the known values of the Recipe over each other and look along the D Scale to the Number of Servings you wish to make and find the amount of the unknown needed material above. Be sure to employ the proper use of scientific notation when needed.

7) Keep in mind, you can determine how many servings you are making by looking in the opposite direction and going from what you have and looking back on the D Scale at how many servings it will make!

Basic Conversions :

1) The Slide Rule is the best conversion system around. This is because this operation is similar to the percentage calculation and the recipe calculation above.

2) Here, in this activity we can convert fractions to decimals or vice versa. We can also find equal fractions as well.

3) Also, if we need to convert one unit into another (inches into feet, feet into yards, ounces into pounds, inches into centimeters, and even complex ones like mph into kph) this is considered here.

4) In the case of fractions simply set the C Scale as the Numerator and the D Scale as the Denominator. Reading along the C Scale one finds all of the similar ratios.

5) (For example, if 2 on C is over 3 on D one finds 4 on C over 6 on D, and 6 on C is over 9 on D, et al. –

6) Not only that, but above the D Scale Index we find the decimal equivalent of the fraction, 0.66)

7) This means for any fraction or decimal we can find the other easily.

8) What is needed is first a ratio of what is known, say one unit over another. (For example 1 inch = 2.54 cm, 8 ounces = 1 cup, 1 foot = 12 inches)

9) Set up this ratio on the slide rule. It can be in either order where one is the C Scale value and the other is the D Scale value.

10) Now look for the other known value along the line that is known. On the opposite Scale will be the sought out answer.

11) (For example if Inches are on the C scale and Centimeters are on the D Scale and we wish to know how many centimeters are 4.5 inches, we read along the C Scale to 4.5 and find the answer below on the D Scale).

12) What this implies is that with a simple list of conversion factors, one can readily perform rapid calculations.

13) Conversions can include currency, one set of units into another, and the like.

14) For Currency conversions : 1 unit (say the dollar) equals N units of another currency. Set the Left Unit (1) over the equivalent on the D Scale. Reading along C is the number of dollars and below is the number of corresponding units in the other currency!

15) Another fact is this : Recall our calculations previously for miles per gallon, average speed, and cost per unit ounce.

16) Notice that each of these is a 3-Variable Formula where :

17)

18) What this means is that if the ratio is known (X & Y) then the answer is always found above the D Scale Index (N).

19) Also if we know the outcome value (N), we can estimate answers more readily.

20) (For example if our average speed is 45 mph, how long will it take to travel 280 miles. With 45 on C over 1 on D, read along C to 280 and below it on D is the answer : 6.2 hours)

21) This is how any and all 3-Variable equations can be treated!

22) Some of the basic set of 3-Variable functions needed are as follows :

$$\text{Rate} = \frac{\text{Cost}}{\text{\# Items, etc}} \qquad v = \frac{\Delta d}{\Delta t} \quad a = \frac{\Delta v}{\Delta t} \quad V = I*R \qquad P = V*I$$

23) We have used the first 2 here in this activity amongst others (area, etc) while the last 2 are Ohm's Law concerning voltage, current & resistance, and the final one is electrical power. The middle one is the formula definition of acceleration.

24) Note that there are many more and this is a small list. The Activities have these to learn from and use. Some involve the amount of food one eats and the graphical breakdown of these items. Another Activity calculates the amount of energy used and power exerted in exercising.

25) The key to this is your use of the Slide Rule as the means to connect math to the real world.

26) The Proportion Formula can be used in a number of cases (see other Activities involving its use) such as in the height of an object or distance to it.

27) For example if you know your height and measure the height of your shadow on a given sunny day, and measure the height of a shadow of another thing (tree, light pole, house) you can determine its height.

$$\frac{\text{Your Height}}{\text{Size of Your Shadow}} = \frac{\text{Height of Object}}{\text{Size of Shadow}}$$

Home Economics - Simple Interest on a Loan or Savings Account

1) This calculation is a quick look at the Basic Interest Equation and not the more complex one of being compounded daily or with some other time value.

2) Here, the Principal (P) is known, and perhaps the Interest Rate (R), along with the time period (T).

3) From these variables we can find the amount of accrued interest for a total time period (I).

4) This is a multiple step problem, so follow the directions.

5) First place P on the C Scale over the Index of the D Scale.

6) Read along the D Scale to R and find the answer to this on the C Scale, which we will call 'V'.

7) Now move 'V' to the D Scale Index and again read along the D Scale to the variable T.

8) Above T find the Answer I now on the C Scale.

9) Be sure to rearrange the formula as needed if solving for some other variable.

I = P*R*T

More Complex Loan Equation Calculations

1) This next excursion is the maximum use of math for this activity and requires concentration. Here we are going to examine the compounded interest rate on a loan or savings account.

2) First look at the equation :

$$A = P*(1+\frac{r}{n})^{n*t}$$

3) Here : P = Principal value, r = Annual nominal interest rate expressed as a decimal value, t = the number of years, n = number of times the interest is compounded per year, and A = Total Amount

4) All of the variables are known here except A. We want to determine the total amount owed or earned from the Principal at these given rates and conditions.

5) First compute with your slide rule r / n by using 'r' on the C Scale over 'n' on the D Scale. Call this X. Jot it down for reference.

6) Add 1 to X and multiply this by P on the slide rule.

7) For example place P on the C Scale over the D Scale Index and read along the D Scale to (1+X) and find the answer, Z, on the C Scale above.

8) Next multiply n*t on the slide rule. Call this M and jot it down.

9) Rewrite Z in Scientific Notation.

10) Reading Z (see step 7) on the D Scale, find the log (Z) on the L Scale. Recognize that this is the characteristic and the mantissa is the exponent from the Scientific Notation value. This new value is Q.

11) Now multiply this value, Q, by M (step 8) and find the value W.

12) Look up the characteristic (all of the numbers past the decimal) of W on the L Scale and find its corresponding value on the D Scale.

13) The decimal place is determined by the mantissa of W (the numeric value before the decimal).

14) With proper placement of the decimal, the value read on the D Scale is the Total Amount, A.

15) This takes time to do and to understand, but with patience, practice, and determination, you can succeed. As Slide Rulers say, Keep On Computing!

Activity #2
Exploring Waves with a Spring
Grade Level : High School
Math Level : Calculating

Speed of a Wave on a Spring

Before exploring any of the Wave Activities, it may be best to read the Story of Waves in Ch V to have an overview of the topic.

The Activity Description

In the Activity there are two methods of approach for measuring the speed of waves in a spring (best to use a Slinky or some other item similar to it). One method involves tension in the spring (basically stretching it and holding it in place as best one can) while the other is to vibrate the spring so as to attain natural resonant frequencies.

Many objects have frequencies associated with them, hence this is why we make instruments out of different materials and they have certain shapes. All of this involves the wave behavior of what is referred to as Resonance. First, there is the concept of natural frequency. **Natural Frequency** is the frequency at which an elastic object, once energized, will vibrate. Look at guitar strings on a guitar sometime. They have different thicknesses. When plucked each vibrates at a different frequency than others of different thicknesses. Each has its own characteristic frequency. The frequency will depend with mass, length, and tension on the string. This is for all stringed items, guitars, violins, pianos, and harps for example. Not only do strings do this, but many other objects have natural frequencies associated with them. Items such as drums, bells, tuning forks, and the like all vibrate at particular frequencies depending on their material, shape, and other factors. This will be employed in our lab in the use of the spring.

Resonant Frequency is the minimum energy required to continue the vibration at that frequency. Resonance is a phenomenon that occurs when the frequency of forced vibrations on an object matches the object's natural frequency. Acoustic Resonance is the tendency of an acoustic system to absorb more energy when it is forced or driven at a frequency that matches one of its own natural frequencies of vibration (i.e. its resonance frequency).

By vibrating the spring in a back-and-forth manner with increasing amounts of effort (energy), the majority of the time the spring seems to have no 'rhythm' to it, but at certain levels of energy, the spring will vibrate back-and-forth at a resonant frequency equal to one of its natural frequencies.
The goals of this Activity are to measure and determine the wave speed in a spring, both longitudinal and transverse waves. Next these will be compared to a formula for wave speed in a spring when stretched to a given length and under a measured amount of tension.

Overall Purpose : To create and measure from a mechanical wave the distance and time of pulses on a spring to determine wave speed and to determine the role in tension on the spring as it relates to wave speed.

Activity I :
Purpose : To measure a compression wave speed on a stretched spring of measured tension and compare the measured wave speed to the predicted wave speed due to tension.

Activity II :
Purpose : To measure a transverse wave speed on a spring for comparison to other wave speeds determined on the spring.

Activity III :
Purpose : To use the property of standing waves for a spring to determine the speed of the waves traveling the spring.

Materials :

- Large Spring (Slinky or similar type works best),
- Timer,
- Long Measuring Tape,
- Tension Scale,
- Mass Scale (may be needed, but can use the Tension Scale),
- Tiled floor (best), (wood will work too),
- Graph Paper,
- Goggles,
- Slide Rule

Procedure :

For All Activities :

1) It is best to have 2 people for this activity since one will hold the spring, but it can be done with one as long as one end of the spring is secured in some manner.
2) Have adult permission and supervision. Wear safety goggles. Always also check the system for the spring so that it is held and remains secure throughout the activity.
3) For all activities, the spring is stretched out along the tiled or smooth floor. In the first activity, the length will vary with each trial. For the other two activities, the length is held constant (in fact it is best to have it the same for each of these exercises – perhaps 1.5 or 2.0 m)
4) Along the spring have the measuring tape open so as to be able to determine the length of the stretched spring.

Activity I : Longitudinal (Compression) Wave Speed Determination

1) Follow the directions for all activities above.
2) Use the scale and measure the mass of the spring (M).
3) In this activity, the spring will go from a shorter length to its maximum length (L) (the range could be 1.0 m, 1.5 m, 2.0 m or some other combination depending on the spring and space to do this activity).
4) With the spring set up use the tension scale attached to the end to measure the amount of tension required to hold the spring at this length. Record this value (F_T).
5) Note : Do not leave the scale attached to the system. You need to find a way to attach it – Books holding a plastic bar between them across the spring may work. In any case, be sure it is secure and be safe – wear goggles and do not stand close. Also do not use too much tension – if the spring were to come off its system, it should not be under so much tension so as to be a danger.
6) With the spring set up, send a compression pulse along the spring. It should bounce back from the 'fixed' end (someone else holding it or it is held in place by some means).
7) Do a practice run to note how long this takes. Use a timer.
8) Now conduct the actual trials and time how long it takes a wave pulse to travel the length and back on the spring. (t)
9) Do this 3 times for each length and average the values.
10) Redo this for 2 other lengths, each with 3 trials per length and average them. Be sure to record the new distance (L) and the new tension force (F_T).
11) Calculate Average Speed of the Wave from the distance and times.
12) Calculate the predicted speed from the Wave Speed due to Tension equation.
13) Compare results.

Activity II : Transverse Wave Speed Determination

1) Have the spring held at a constant distance for all trials in this part of the activity. (For example 1.5 m or 2.0 m). (L)
2) Send a Transverse Wave along the spring (that is, snap it) and test time the wave and its echo return wave.
3) Now conduct this trial 3 times – record the 3 times for the wave to travel the total distance (2*L) and average them. (t)
4) Use the Average Speed formula and calculate the speed of the wave.
5) If you want, try other lengths (for example the 3 distances used in Activity 1 here) for comparison.
6) Like Activity 1, be sure not to have too much tension, be sure to wear safety goggles, and be sure to secure the spring and monitor it.

Activity III : Standing Transverse Wave Speed Determination

1) For this activity, the spring must be held firmly in place like the other cases. Here, One person is merely the anchor and not moving, while the other is the agitator and energizes the spring with a back and forth motion to make the spring oscillate.
2) When the spring has a half wave resonance (you see one crest only) which then bounces back and becomes a trough, begin counting 10 oscillations which are timed. (t)
3) Note that one oscillation is a complete cycle, in the first frequency it will go from peak to trough to peak again – this is one. In essence it has a holding pattern that reverses itself and then returns to this pattern.
4) Do this for the next two fundamental frequencies for the spring at this length. The second pattern should require enough energy to have a complete wave. The third pattern will have 1.5 waves at one time.
5) You will have to work at it to find the right amount of energy for each case. Be sure to wear safety goggles.
6) From the data, determine the Period and the Frequency of the Wave.
7) Use the Wavelength and either the Period or Frequency in the Wave Equation and determine the Speed of the Wave.
8) Time permitting, try another length of the spring and redo the exercise.
9) From all 3 exercises compare the wave speeds and note similarities and differences.

Data :

Activity I : Longitudinal (Compression) Wave Speed Determination

Mass of Spring (M) : _____ (kg)

For each Distance, measure the spring Tension and take the average speed of the wave pulse for 3 trials. Complete this 3 times (that is 3 different distances of stretched spring)

Each trial has its own length, tension force, and speed. Each will have its own speed. Compare this average speed to the computed speed using the wave speed due to tension force equation.

Distance of Spring [L] : _____ (m)

Tension of Spring [F_T] : _____ (N)

Trial	[D] Distance for Wave (2*L) travel	[t] Time for Wave Pulse Travel (s)
1A		
1B		
1C		

Activity II : Transverse Wave Speed Determination

For the given distance (L), maintain this for each trial. Do at least 3 trials
and average them and then determine wave speed.
Note : Here you are using a Transverse Wave (pluck the spring
perpendicular to its length to send a pulse along it).

Distance of Spring [L] : _____ (m)

Trial	[D] Distance for Wave (2*L) travel	[t] Time for Wave Pulse Travel (s)
1		
2		
3		

Activity III : Standing Transverse Wave Speed Determination

Do this entire session twice, but maintain the stretched length of the
system. Then average the corresponding results from each data set.

Stretched Length of System [L] : _____ (m)

Trial	Wave Description	[λ] Wavelength (m)	[t] Time for 10 Oscillations (s)
1	2L		
2	L		
3	$\frac{2}{3}$L		

Trial	[T] Period (s)	[f] Frequency (s^{-1} , Hz)
1		
2		
3		

Calculations :

Be sure to use your Slide Rule!

$$v = \frac{\text{distance traveled}}{\text{time for traveled distance}}$$

Average Speed Formulae (know which variables to use)

$$v = \frac{d}{t}$$

Wave Equation :

$$v = \lambda * f$$

Wave Speed due to Tension Equations :

$$\mu = \frac{m}{L}$$

$$v = \left(\frac{F_T}{\mu} \right)^{1/2}$$

μ is mass per unit length (stretched distance), L is stretched distance, F_T is the tension force in the spring for that distance stretched

Frequency-Period Relation :

$$f = \frac{1}{T}$$

Frequency & Period :

$$f = \frac{\text{Number of Oscillations (n)}}{\text{Time for oscillations (t)}}$$

$$T = \frac{\text{Time for Oscillations (t)}}{\text{Number of Oscillations (n)}}$$

Conclusion :

In Activity I, examine how the speed changes (or not) as the length and hence tension on the spring changes. Also, how do these average speed calculations compare to the wave speed predicted by the tension force equation for speed?

From Activity II, how do the wave speeds compare in the trials to each other? How does this transverse wave speed compare to the compression wave speed in Activity I? How does the transverse wave speed compare to the predicted wave speed from the tension force equation for speed?

In Activity III, how do the wave speeds compare to each other for each of the standing wave patterns? How do these transverse wave speeds compare to the compression wave speeds of Activity I and the transverse waves of Activity II?

Activity #3
Several Pendulum Characteristics Explorations
Grade Level : Middle School
Math Level : Calculating

The Period of the Pendulum and the Slide Rule Activity

How do Pendulums end up in a book about Waves, one might ask? One of the most common descriptions for wave behavior is the use of a Sine wave which can be readily found with a pendulum. Imagine a cone-shaped pendulum with sand that can come out of the point, or better yet one with a pen stylus on its end. Have a paper scroll move at a right angle in a given direction beneath the oscillating pendulum so that it leaves a sand trail or scribes a line. It will be a sinusoidal wave pattern – the classic wave pattern we are all accustomed to. Mathematically wave formulae (sine wave functions) are often the basis of wave motions.

A simple Pendulum is a given mass at the end of a 'rod' (chain, string, solid rod) that is allowed to swing back and forth freely with respect to a fixed point at the end opposite the mass. Gravity is the main driving force for this system.

A simple natural question arises from this : What factor or factors affects the period of oscillation (the amount of time to swing back and forth)? Is it the mass, the type of mass, the type or length of the rod or something altogether different than these ?

This question came to Galileo and was recounted in his biography from accounts of him watching the swaying of bronze chandelier in the cathedral of Pisa and timing it with his own pulse (note that pocket watches did not yet exist). In his book, Two New Sciences, Galileo concludes from his experiments on pendulums that pendulums of the same length independent of mass are isochronous (taking the same amount of time). His mathematical analysis found that the square of the period varied with the length of the pendulum. He also thought that it was independent of amplitude (the size of the swing) which is not true and later investigations show that Galileo's conclusions are true for angles $15°$ and smaller.

These conclusions follow from Galileo's understanding of how the acceleration due to gravity does not depend on mass (all masses change speed at the same rate in a state of free fall).

Pendulums have uses too. By the 1700s, parts for clocks can be made small enough to develop clocks that are mechanically powered by masses that fall and are kept in time by a swaying pendulum, or as they are more commonly called today 'grandfather clocks'.

A pendulum seen edge-on is akin to an orbiting satellite about its parent planet or a planet about its star. This type of back-and-forth motion has a name (when the restoring force causing the motion varies with the distance from equilibrium) and it is called Simple Harmonic Motion. A pendulum swaying back and forth is similar to a mass attached to a spring moving back and forth.

Pendulums are a very good examination of conservation of energy. When at their highest points, the pendulum has zero velocity, hence zero kinetic energy. But it still has energy in the form of gravitational potential energy which is at its maximum since this increases with height. As it falls, the potential energy decreases, but the energy does not disappear. This turns into kinetic energy which manifests itself as a changing and increasing velocity. At its lowest point of the swing, which is the equilibrium position, it achieves maximum velocity hence maximum kinetic energy while having no relative potential energy. This also explains why the pendulum climbs back to its original height (as noted by Galileo too in this and studies of inclines). This is because energy cannot be created nor destroyed so it turns back into potential energy. With each swing, there is some energy loss due to friction (an idea described by Galileo) so in time the displacement decreases until it stops oscillating and the energy has all turned to heat and sound.

This activity explores the factors that could affect a pendulum in order to reach the same conclusions that Galileo did in 1638.

Activity I : Mass & Pendulum Period
Purpose : To investigate and mathematically uncover the relation of the period of a pendulum to its mass.

Activity II : Mass Type & Pendulum Period
Purpose : To investigate and mathematically uncover the relation of the period of a pendulum to its type of mass (solid or liquid).

Activity III : Rod Type & Pendulum Period
Purpose : To investigate and mathematically uncover the relation of the period of a pendulum to its type of pendulum rod (rigid or flexible).

Activity IV : Pendulum Rod Length & Pendulum Period
Purpose : To investigate and mathematically uncover the relation of the period of a pendulum to the length of a pendulum.

Materials :

- String,
- Nuts and bolt (eye form is best − large for both),
- Meter Stick or Measuring Tape,
- Mass Scale,
- 16 oz bottle,
- Water,
- Scissors,
- Wooden Dowel Rod (solid rod in Activity III),
- Eye Hook,
- wooden rod (1" x 1" x 2 ft),
- Chairs (2),
- Protractor,
- Timer,
- Graph Paper,
- Slide Rule

Preparation before Activity Procedures :

1) Always have parental permission and help in Activities. Always employ safety in dealing with all materials.
2) Note that each of the data tables provided is minimal and needs more entries for each of the trials – a good number of trials is at least 3, but 5 is recommended.
3) Use eye bolts and nuts that can attach easily that are as large as possible. These can act as incremental masses by name instead of amount if no scale is present. The best choice is with a scale.
4) With a scale present, mass out the eye bolt individually followed by additional masses (attached nuts) one at a time. Keep these in order of use in the activity.
5) Have 2 conventional chairs with wood/plastic/metal bodies and flat top backs to act as platform to operate from. Place rectangular board across it and screw into it the hook which can be an open eye hook hanging down for pendulum rod attachment. Be sure to have board between chairs secured (tape, other ?).
6) If doing the solid rod pendulum, a closed eye hook on one end is recommended as this can readily attach to the system set up for the pendulums as noted in the prior steps.
7) For all pendulum situations, regardless of type, length, or mass all will be drawn back no more than 15°. To do this use the Protractor held beneath the board with the eyehook and the rod system (either string or rod) so that when drawn back the pendulum rod does not move more than a displacement of 15°.
8) Each of the Activities recommends 10 oscillations, since this makes the math much easier, but you do not have to use this number – but whatever number of oscillations are chosen, be consistent with each of the trials in a given Activity and it is best for comparisons across Activities to have the same number of oscillations. Note that this can be an Activity unto itself – one where the number of oscillations is small such as 4 or 5 and one where it is larger such as 12 or 13 for comparison results – plus it gives you greater practice in the use of the slide rule.
9) For the solid rod to be used attach a small eyehook in the end carefully.

Procedure :

Activity I : Mass & Pendulum Period

1) For this activity, string will be the choice of pendulum rod type. Length of system can be a choice and should be between 20-40 cm. Each end of the string has loops tied in it for attachment to the hooks at either end.
2) The masses to be used are solid and will be the eyebolt and nuts. Read the preparation steps above. For setting up the pendulum system and having the masses ready to go.
3) Attach the string to the eyehook and attach the eyehook. Measure from the top attachment to the bottom of the eyehook. Record this in the data table.
4) Record the 1st mass as the eyehook mass in trial #1.
5) For each trial draw the pendulum back the same angle regardless of mass used.
6) When released start the timer.
7) Let the pendulum oscillate back and forth. One complete cycle is one oscillation or period.
8) Count 10 of these and then stop the timer.
9) Record the time of the timer in the data table.
10) Add more mass and repeat the cycle process of step 5.

Activity II : Mass Type & Pendulum Period

1) The type of rod to be used is string and the length will be constant for all measures in this activity (choose a given length of string between 20-40 cm).
2) The mass to be used is water and will be held by a pop bottle attached to the string. Attach the string so that the bottle is open or can be opened.
3) Begin with a measured amount of water mass either via scale measurement of water in measuring cups on a scale or simply use measuring cups and calculate the mass.
4) Use incremental amounts of water starting with ¼ cup and add ¼ cup increments to the system.
5) Draw back the pendulum the same amount of distance each time and let it swing 10 oscillations timing them with the timer. Record the time needed for each trial.
6) For each trial increase the amount of mass in water.

Activity III : Rod Type & Pendulum Period

1) In this Activity you must decide how to attach the given mass (which is constant for this activity) to the solid rod – you can consider duct tape – a possibility is to tape one of the bolts with the head at the bottom of the solid rod and the bolt points straight down from the rod so that nuts can be attached. Decide on the amount to be used.

2) In this activity, the mass is held constant as is the length of the pendulum rod. The string with the mass must be the same length as the mass on the solid rod.
3) There are only trials of the string pendulum and the solid rod pendulum systems. They can be done more than once to verify the overall results.
4) Always draw the pendulum back the same initial distance and it is less than 15°.
5) As with the other activities, let the pendulum swing or oscillate 10 complete times and record the amount of time for this.

Activity IV : Pendulum Rod Length & Pendulum Period

1) Choose a solid mass that acts as a constant for all trials in this case where the question of pendulum rod length is being investigated. The nuts and bolts are a good choice in this Activity.
2) Measure out approximately 50-60 cm of string and wind it around the main support rod for the pendulum system so that the initial distance is shortest.
3) Begin with the shortest distance of string for your trials, between 10-15 cm with the mass attached.
4) As with all trials draw back a short yet constant angular distance of less than 15° for each trial.
5) Let each pendulum length oscillate 10 complete times and record the time for this event.
6) Unwind more and more of the pendulum string so that the length of the pendulum increases in some sort of incremental fashion for each trial. Then follow steps 4 & 5 until all trials are done.

Data :

Activity I : Effect of Mass on Pendulum Period (solid bob)

Type of Pendulum Rod : _____

Length of Pendulum Rod : _____

Trial	Mass (unit or g)	Time (t) for 10 oscillations (s)	Period (T) for 1 oscillation (s)
1			
2			

Activity II : Effect of Type of Mass on Pendulum Period (liquid bob)

Length of Pendulum Rod : _____

Trial	Mass (unit or g)	Time (t) for 10 oscillations (s)	Period (T) for 1 oscillation (s)
1			
2			

Activity III : Effect of Pendulum Rod Type on Pendulum Period (rigid & flexible)

Mass of Pendulum Bob Used : _____

Type of Pendulum Rod Used : _____

Length of Pendulum Rod Used : _____

Trial	Length [L](cm)	Time for 10 oscillations (s)	Period for 1 oscillation (s)
1			
2			

Activity IV : Effect of Pendulum Length on Pendulum Period (string)

Type of Pendulum Bob : _____

Mass of Pendulum Bob : _____

Trial	Length [L](cm)	Time for 10 oscillations (s)	Period for 1 oscillation (s)
1			
2			

Calculations :

Be sure to use a Slide Rule !

Watch your calculations when conversions, such as centimeters to meters are needed.

Procedure :

1) It is best to use the C & D scales of the Slide Rule for the calculations. If squaring is needed be sure to employ the A or B scales.
2) For each activity, determine the period of the pendulum by using the formula for period (T).
3) For Activities I & II, graph Period vs. Mass and draw a best fit line. Determine its slope. (It should be zero – what does this indicate?)
4) For Activity 4 graph Period vs. Length.
5) Recreate the Table for Activity IV and take the log value for the Period and the Length.
6) Graph the Log (Period) vs. Log (Length) and draw a best fit line and find the slope of this line to determine the exponential relation between Period and Length. (It should be: $L \sim T^2$). (The slope ideally is 2). Note the log value is found on the L scale of the slide rule.
7) One way to verify the relation is to graph Period^2 vs. Length which should turn out to be a straight line.
8) Use the ideal formula for period as determined by length and calculate what your period values should have been as compared to what you measured them as. Note that the ideal formula may vary – for most pendulums the simple formula suffices, but if the rod has a large mass as compared to the bob, then the complex formula for period should be used.

Calculations :

Be sure to use your Slide Rule!

Information :

      1 mL water = 1 cc water = 1 g water

Formulae :

$$T = \frac{t}{10}$$

      (Note : Assumes using 10 oscillations – instead of 10 put your number if it is different)

$g = 9.8 \ m/s^2$

Formula for simple pendulum (mass concentrated in bob) :

$$T = 2*\pi*(L/g)^{1/2}$$

$$L = \frac{g*T^2}{4*\pi^2}$$

Formula for Complex pendulum (mass of rod is considerable as compared to bob) :

$$T = 2*\pi*((2*L)/(3*g))^{1/2}$$

$$L = \frac{2*g*T^2}{3*4*\pi^2}$$

Conclusion :

Examine your results to find which if any of the variables have affected the period of a pendulum. From the reading, results, and calculations it should be found that the period is related to the length of the pendulum.

It might be interesting to test the idea of whether the complex period formula is needed or not – have a rigid rod pendulum system activity where the mass of the bob is small as compared to the rod and one where the mass of the bob is considerable as compared to the rod and see if these results are similar or different.

Side Note : When two variables are graphed opposite each other and there is either a random pattern or a zero slope, then this indicates no relation to the variables. If instead the two graphed variables show a regular change (linear or not) there is an indication of a relation between them.

Summary :

For the teacher, parent or interested student :
All of the activities explore the relation of period to length and mass, but
only for small angles of angular displacement. Clearly this is a hypothesis
to explore as well. Choose a length and mass for a given set of trials,
then change the angle at which the pendulum starts. For example, start
at 10°, then 20°, then 40°, then 60°. Graph these results to see if there
are any variations.

An important idea arises in this Activity. Note that several of the activities
in it show no relation of the variables. One might think why go down this
road at all and only pursue the other possibilities? Simple – Just because
the idea from the hypothesis as extended from known ideas in science
seems to indicate one answer (the length of the pendulum) does not
mean that the question concerning the other variables (such as mass)
have been answered. A failed experiment is just as important as a
successful one. The process is about being true to the numbers and
realizing what they are telling us.

Also, try other challenging ideas : make a pendulum with a certain Period,
for example 1.0 seconds. Compare your outcome to the predicted value
found in the Pendulum Period equation.

Interesting Use of the Pendulum :
As you found in your investigation, the period of the pendulum depends
on length of the pendulum. With this idea in mind, one can construct a
1 s pendulum as noted above, but can test the formula from the activity
for various lengths of pendulums and the period that is predicted for it.
Construct and test these pendulums. Be sure to run several trials and
take the average of the values. Also note, the shorter the pendulum, the
greater the potential for error due to the fact that the period becomes
shorter, hence your reflexes at measuring its period will vary
considerably.

A Personal Stopwatch :
An interesting use of these pendulums is to use them when there is no
stopwatch available. Though one might not have the exact time for the
pendulum known (in our case, if tested we do), the pendulum can be
used in place of a timer in any or all of the activities in the book. Note
that the precision of the tool decreases considerably. In the case of using
it, just count the whole number of swings that the pendulum goes
through for a given timing. Since for each trial it has the same value, all
of our measurements for that activity using our makeshift stopwatch are
valid!

Activity #4
Determining the Speed of Sound in Air
Grade Level : Middle School
Math Level : Calculating

Sound is a mechanical wave that results from the vibration of the medium that has the sound wave traveling through it. In air it is a longitudinal wave form and has areas of higher and lower density hence pressure called compressions and rarefactions. Though the most common is air, sound waves can travel through liquids, like water, and through solids, such as metals and the like.

The transmission of sound is best visualized via a model wherein the atoms of the medium are like marbles attached to each other by springs. The sound wave is a push on atoms in one part of the system so as to compress the springs. This transmits energy to the neighboring marbles, which move due to the increase in energy and then compress the springs to the next set of marbles and so on until the wave propagates across the space. Note it is not the atoms themselves that travel from one point to another, but the wave through them. So, when someone speaks to you, the atoms nearest them are still there, but the sound wave goes through the medium to your ears which pick up the sound.

This also explains why sound spreads out as it travels, like light or gravity. With distance, its intensity decreases as the inverse square of the distance involved. (See inverse-square law activity #11).

Sound Intensity is proportional to the square of the amplitude of the wave involved. It is related to but not the same as loudness of sound. Loudness varies with the logarithm of intensity (i.e. powers of 10). It is measured in **decibels (dB)** after Alexander Graham Bell, the inventor of the phone who also did work in the science of hearing and hearing impaired people. In the decibel system, 0 (zero) is the threshold of hearing and every 10 dB is an increase by a factor of 10 in intensity. A sound at 10 dB is 10 x more intense than at zero dB. A sound at 20 dB is 10x10 or 100 x as intense as a sound at 0 dB. The same is true for the difference between 50 dB and 70 dB. The difference of 20 dB means an intensity factor of 100 x.

$$I(dB) = 10\log_{10}(-)$$

Source of Sound	Level (dB)
Jet engine at 30 m	140
Threshold of pain	120
Average factory	90
Busy street traffic	70
Normal speech	60
Close whisper	20
Normal breathing	10
Hearing threshold	0

Also the marble with springs model (see above for description) explains how waves bounce off of a material (reflection).

All Mechanical Waves in a given situation under constant conditions will travel at the same speed regardless of their frequency. For example sound waves in air at the same temperature and other atmospheric conditions, will travel at the same rate of speed (about 331.5 m/s at 0° C). It turns out that temperature is one of the more important factors for wave speed. The warmer the air, the faster the wave speed will be. For example at 20° C, the speed of sound is about 343 m/s (1,125 ft/s which is 1,236 km/hr or 768 mph). These noted figures are about 1 mile in 5 seconds or 1 kilometer in 3 seconds. Hence this is where we get the idea of counting so many seconds after seeing lightning and then hearing thunder so as to determine the distance to the storm. When it comes to temperature there is even a quick relation to give a general value for it, which we will use in the Activity and use it as the basis of the actual value (though it is not).

$$v = 331.5 \text{ m/s} + 0.6*T$$

An interesting fact is that waves in solids are faster than in liquids and these are faster than waves in gases (on the average). The speed of sound in air is about 331.5 m/s, while in helium it is 972 m/s. The speed of sound in water is about 1490 m/s yet in sea water (since there is salt and other minerals dissolved in it) it is 1530 m/s. The speed of sound in solid copper is 3560 m/s, while in solid aluminum it is 5100 m/s and in solid iron it is 5130 m/s. (Note that this is primarily true for solids that are crystalline in nature).

Note that independent of the frequency or even amplitude of the sound, under the same conditions it will travel at the same speed in the air. For example, have two instruments, like a flute and a bass drum at equal distances from you and strike a note at the same time. The sound waves will arrive to your ears at the same time.

The most common phrase for the speed of sound in air is **Mach Speed**. The Mach Number is the ratio of the speed of the item in question to the speed of sound. If the ratio is 1, then the object is traveling at the speed of sound! Therefore, mach 2 means moving at twice the speed of sound. Note that it is not only temperature in this case, since the aircraft doing this are at much higher altitudes where pressures are different, so there is a formula for the Mach Number with respect to pressure.

The **sonic boom** that accompanies airplanes moving at the speed of sound or greater comes from the overlapping of the waves since the craft is moving faster than the sound waves given off. These overlapping waves create the boom we hear. This occurs after the craft has passed since like a boat moving through the water the wave of the craft through the air follow the craft. Other more slowly moving craft have their sound waves move at a constant speed hence we hear a continuous noise instead of the intense boom.

Probably the first speed of sound to be sought in science was the speed of sound in air. In fact, when the phrase speed of sound is noted, by implication it is about air. One of the most notable persons looking into it was none other than Sir Isaac Newton. In his second book of the Principia he notes his derived value for the speed of sound which was about 16% off the present day value, but later corrected by Laplace by examining the information. Further scientific study of sound leads to the famed **Doppler Effect** where a moving object emitting sound has a higher frequency than the at-rest transmitting frequency of the sound source according to an observer in front of the oncoming sound source. When it passes, the frequency becomes less than the original frequency. This was originally described for sound of a train whistle. This shift, due to the motion of the sound source, was transferred to light and is the basis of understanding the motion of objects, planets, stars, and galaxies in astronomy and was the first realized evidence for the Big Bang by Hubble in the 1920s.

The earliest efforts at measuring the speed of sound would be to have some distance between the two scientists. Each with synchronized pocket watches one fire a gun while the other noted the puff of smoke through a small telescope and then times how long it takes for the sound to reach that point. With distance and time known, the average speed can readily be calculated.

There is an easier way and on a smaller scale. It involves the wave behavior of resonance. First, there is the concept of natural frequency. Natural Frequency is the frequency at which an elastic object, once energized, will vibrate. Look at guitar strings on a guitar sometime. They have different thicknesses. When plucked each vibrates at a different frequency than others of different thicknesses. Each has its own characteristic frequency. The frequency will depend with mass, length, and tension on the string. This is for all stringed items, guitars, violins, pianos, and harps for example. Not only do strings do this, but many other objects have natural frequencies associated with them. Items such as drums, bells, tuning forks, and the like all vibrate at particular frequencies depending on their material, shape, and other factors.

Resonant Frequency is the minimum energy required to continue the vibration at that frequency. **Resonance** is a phenomenon that occurs when the frequency of forced vibrations on an object matches the object's natural frequency. This is like the case of two swings swaying similarly when one of them is set in motion on a swing set. Acoustic Resonance is the tendency of an acoustic system to absorb more energy when it is forced or driven at a frequency that matches one of its own natural frequencies of vibration (i.e. its resonance frequency).

Tubes can have resonant frequencies too. Their resonant frequencies are related to the length of the tube, the shape of the tube, and whether it has closed or open ends. Many instruments are pipes such as lip-reed instruments like clarinets, saxophones, and oboes, open pipe types like flutes and piccolos, and even organs with closed and open pipe systems.

In the open pipe case when a sound is played that is the same resonant frequency of the pipe, the first frequency will be ½ wave, while the second resonant frequency will be one full wave. With each resonant frequency, it will add one-half wave.

$$f = \frac{*}{*(\quad . \quad)}$$

n here is a positive integer 1,2,3...

In the closed pipe case, when the sound is played and the closed end pipe responds with its resonant frequency the first resonance frequency is ¼ wave, the next is ½ wave more so it is ¾ wave and so on.

$$f = \frac{*}{*(\quad . \quad)}$$

f is frequency, n is an odd number 1,3,5, v is wave speed,
L is pipe length, d is pipe diameter

Our activity will take advantage of resonance in a closed pipe. We will take a plastic pipe and have it inside of a vessel (the most commonly used is a 1 L graduated cylinder, but these can be expensive - hence, why not use a tall plastic pitcher instead! If this idea is impossible, then have an outer plastic pipe for the inner one). In the outer container   (graduated cylinder, pitcher, or pipe) there is water which acts as the stopper for the pipe. As the inner pipe is drawn up there is larger and larger length of pipe exposed. Now strike a tuning fork above it and slowly draw the pipe up. As the pipe is drawn up the sound intensifies since it is bouncing off the end and the air in the pipe will resonate since it is at the right length that corresponds to a fraction of the wave size. Notice that the formula has the speed of the wave in it. We know the frequency being used (marked on the tuning fork) and we know the wavelength (derived from the pipe length). We use the wave equation and hence we have determined the speed of sound in air right where we stand with two pipes, a measuring tape, water, and a tuning fork – all due to the property of resonance!

Purpose : To use the wave property of resonance to determine the speed of sound in Air.

Materials :

- Tall Pitcher (the taller the better) or tall plastic container – height needs to be about 35-40 cm minimum,
- Water,
- Tuning Forks set (255 Hz, 320 Hz, 384 Hz, 512 Hz, et al),
- 1.0" diameter pvc pipe cut to 10 cm taller than the container used,
- Measuring Tape,
- Thermometer,
- Slide Rule

Note : If not able to find a tall pitcher, there is the basic need of a tall water containing item. A tall 1 L graduated cylinder is possible, but glass is costly – the plastic ones are less expensive. Another alternative is to have a 1.25" pvc pipe where the 1.0" piece used in the materials fits in it. With this you have to also have a cap for the end of the 1.25" piece glued in place (to make it watertight). – The pitcher is the best choice.

Note : Also if any cutting is done to make any of the needed items – such as the pvc pipe, ALWAYS wear GOGGLES, and Have Parental Permission and Supervision. Always think and work safely.

Procedure :

1) Check the room's temperature and record this value. From this value calculate with a slide rule the 'accepted value' for the speed of sound in air at this time and place. (T)
2) Measure the inner diameter of the Resonance Tube (d) and record this.
3) Follow the Pre-Assembly Procedure where the tubes are first crated and then once created fill the main tube with about 2/3 water (test how much so it does not overflow when placing the Resonance Tube in it).
4) Use each of the Tuning Forks in turn (highest to lowest or vice versa) for each trial. Strike the one you are using (best to use a rubber mallet).
5) Turn the Tuning Fork sideways held above the Resonance Tube. Slowly draw the Resonance Tube upwards so as to increase the length of tube out of the water.
6) Listen carefully and find the exact height when resonance occurs. This happens when there is a high amplitude of forced vibration that vibrates at the same frequency of the tuning fork.
7) Measure the distance from the top of the Resonance pipe to the top of the water with the measuring tape (s). Record the tuning fork frequency (f) and the length of resonance tube out of the water in the table.
8) Hints : Having two people helps the process since holding the tuning fork and moving the tube can be a bit of effort and one could make mistakes.
9) Hints : When pulling the Resonance Tube out it is best to have one had wrapped at the junction of the Resonance Tube and the Main Tube to hold it in place for measurement.
10) Calculate the Speed of Sound in Air from each of the tuning Forks used. Note that you need to first put in the correction factor for the tube diameter. The formula used for speed is the Wave Equation.
11) Calculate the percent error comparing your experimental results to the 'accepted value' for speed of sound in air.

Data :

Temperature of the room [T] : _____ (°C)

Diameter of Resonance Tube (inner tube inside dia) [d] : _____ (m)

Trial	[f] Frequency (Hz)	[s] Resonant Tube Length Measure (m)	Calculated [λ] Wavelength (m)
1			
2			
3			

Calculations :

Be sure to use your Slide Rule!

Speed of Sound based on Air Temperature Formula :

v = 331.5 m/s + 0.6*T

> T is temperature measured in °C

Corrected Length for air column :

L = s + 0.4*d

> s : is the measured air column span in closed tube
> d : is the diameter of the pipe with closed air column

Wavelength based on air-column length data :

λ = 4*L

Wave Equation :

v = λ*f

Percent Error :

$$\%E = \frac{[\text{Experimental Value-Accepted Value}]}{\text{Accepted Value}} * 100\%$$

Conclusion :

First examine your speed of sound in air determinations for each of the chosen frequencies used. Are they similar or not and why? Next treating the speed of sound in air as the one determined by the temperature of air formula, what percent error do your experimental results show as compared to this value?

Note : You should estimate the outcome mentally before you actually do the work with the tuning fork. For example, if the room is 20° C the speed will be about 350 m/s, so this means that a 512 Hz frequency tuning fork will have a value of 68 cm and ¼ of that for the wavelength and the amount of pipe out of the water is 17 cm. Since a 255 Hz tuning fork is about half the frequency of the 512 Hz one, its' length will be twice as long at 136 cm or a ¼ wave at 34 cm.

Activity #5
Using a Microwave to Determine the Speed of Light Activity
Grade Level : High School
Math Level : Calculating

There is considerable information in the Spring Wave Speed Activity and the Snell's Law Activity with regards to Waves in general, so it may be best to start there for an overall understanding of waves in general and electromagnetic waves in particular. Some of the ideas directly related to the topic at hand are noted below in the writing.

Electromagnetic Waves are quite familiar to all of us. We see them in the form of Visible Light, which represents only a small portion of the electromagnetic spectrum. There are many others, however, we hear of often and are used very frequently by humans in their devices and daily lives. We use X-rays to examine the interiors of items, even ourselves when it comes to our bones. We hear of the harm and damage of Ultraviolet Rays and the need for sun block. Infrared rays are used for looking for people, animals, and the like at night. All of radio, TV, satellite phones and the like use a portion of the electromagnetic spectrum to send and receive messages and information which is called collectively Radio Waves. Also in there are Microwaves, not just the device but the waves the device uses to heat water molecules (to get them to vibrate) to heat food and cups of coffee and such.

Other than the fact that all of the waves mentioned are composed of the same parts – a magnetic wave and an electric field wave – there is one other unique feature all Electromagnetic Waves share in common which was found in Maxwell's Equations describing these waves and found to be a cornerstone of the Universe by Einstein and that is : In a vacuum all electromagnetic waves travel at what is classically called 'the Speed of Light'. Though this term is used, it is only because of human study of light first and the others later on. This value is listed in the Calculations section below and will be used in the Activity that follows and is normally seen as 3.0×10^8 m/s or 186,000 mps (that is miles per second not hour as we are normally expecting). If one could send a beam in a circular path around the world, it would circle the globe 7.5 times in one second!

The Microwave uses a device called a magnetron to generate microwaves from electrical energy usage. These waves are contained in the oven chamber even with the perforated screen at front because these holes are smaller than the wavelengths of the microwaves, so they cannot escape.

We might think that the waves in the microwave just randomly bounce around but they do not. The waves set up what is referred to as a Standing Wave Pattern. This is much like taking a slinky spring or a jump rope and making it move with just enough energy so that the item moves in a rhythmic and stable fashion, such as when two friends are spinning the jump rope just right to make what appears to be a half-wave from the side. It appears that the wave is rather stationary and not moving, hence its name. In the case of the jumping rope example, The two people are at what are called nodes and the high point in the middle is the wave crest. What if the two people with the jump rope (or the person moving the slinky spring back and forth) do so even faster.

They can reach another stable speed where a node appears in the middle between them and now there is a complete wave cycle present as viewed from the side.

All of this discussion of standing waves applies to microwaves as well. The waves in the microwave set up a standing pattern. If one had a very small instrument that could measure at different spots across the microwave one would find that microwave energy would rise to a peak, then decrease and peak yet again as one moves across the microwave chamber. The peaks are where the wave crests of the microwaves are while the areas of low to no energy are the nodes. This is why microwaves have a rotating table in them! This way something that is being heated will pass through these regions of peaks and valleys repeatedly. Without rotation there would only be hot spots on the food while the other areas remain rather cold.
We wish to take advantage of these peaks and valleys, so we must disable the rotation of the platter by taking it out as well as the post that the platter sits upon. Now when we go to heat something, which we will do in the Activity, there will be hot and cold spots. We will use something like chocolate or marshmallows (see list below in Activity) and only heat it a very short time to cause just the initial melting of the surface. What we are after is the distance between these hot spots! That distance corresponds to ½ of the wavelength of the microwave wave. We merely have to double this distance and we are half-way to determining the speed of light for these waves.

We have to use what is classically called the Wave Equation.

$$c = \lambda * f$$

It states that the speed of a wave is equal to the product of its wavelength and the wave's frequency. Normally the variable v is used instead of c since this equation is for all wave types (electromagnetic and mechanical alike) but when 'c' is used it is referring to light and all of its associated electromagnetic wave forms since they all travel at a constant speed all the time as Einstein predicted and found.

In our case of our Activity, We have the wavelength, but what of the frequency? This can be found on the manufacturer's sticker for the microwave (be sure to have parental support and help in this) found on the backside or underside of the device. Most microwaves have a frequency of 2450 MHz (2450×10^6 Hz) or one could simply go to the internet and type in the model name and number and with some research uncover the number needed.
As noted this is called the Speed of Light, as this applies to any and all Electromagnetic Waves in a vacuum. One could readily note that there is air in the microwave, and this will technically affect its speed, but not much at all – in fact if properly done your answer should be within a small percentage of error and somewhat reliable to one perhaps even two significant figures.
By the way be sure to follow all the directions below, have parental permission and help, plus use food items that one will consume – don't be wasteful. It is important to note on the side, however, that this Activity is one of the rare ones – in virtually no case does one ever consume anything in a lab exercise, but here the Activity is designed to be safe and allow for that outcome! Enjoy and Explore :)

Purpose : To determine the speed of light from distances between melting spots on food in a microwave due to heating by a microwave standing wave pattern and multiplied by the frequency of those waves

Materials :

- Microwave,
- Microwave-safe baking or casserole dish,
- Food Item to Use : Choose one from : Chocolate bar(s), Chocolate morsels, bag of mini-marshmallows, cheese slices - (note for choice – must be sufficient to span if not cover the dish being used in a uniform layer – part of this depends on the food's use afterwards – good idea, don't waste good food),
- Ruler,
- Toothpicks,
- Goggles,
- Slide Rule

Important Notes :

Note 1 : Must have parental help and permission in this Activity. There are a number of steps which should involve parental help, such as disabling the rotating dish and finding the frequency of the microwave on the back or bottom of the machine (Note – you might find it on the internet with some research on your model), as well as properly handling heating of food items. The key is that you have a microwave where the center post of the rotating mechanism easily comes out – if not, then you cannot – most lift right out. Be sure to read all directions before doing the Activity.

Note 2 : Do not overheat the Food Item being used. The range of time does depend on microwave power and the item chosen, but typically ranges from a handful of seconds up to not longer than about 40 seconds. The key is watching the item while heating and once melting is detected, then stop the process. – Also do not touch it – it may be too hot and you are going to measure it, you don't want to ruin it.

Note 3 : Goggles are noted and should be worn as a good practice for science lab work, but also in the remote chance of the food item being too hot (which if done correctly it should not be) and it bubbles.

Note 4 : Food Choice is by preference and should be done in the knowledge that this food will be consumed or used in some sort of recipe as it is.

Note 5 : The microwave-safe dish is best to be a dish that has a wall so that a rigid ruler can be put across it and one can read looking down at the food beneath the ruler above it spanning the dish rim.

Note 6 : There are others who have written about this lab and have used whipped egg whites or very-buttered bread instead of the list noted above.

Set-Up :
- Be sure to follow all of the Notes above in your procedure
- Before using the microwave it is best to record the frequency of it found on the sticker label (typically on the back of it), f. Record this in your data table. If unable to find it, perhaps try the internet or simply use the suggested value in the table in Calculations as this is the most common value

- In preparing the microwave for use in the lab, you are taking out the turntable and its post (typically the post is plastic and lifts right out) – This is so a dish can sit in there and not rotate when the microwave is on
- Be sure to use a microwave-safe dish and it is best if it covers as much of the microwave floor as possible but also have a span that can be spanned by a ruler that can rest on its rim
- In terms of the ruler, it is best to use a solid one (that is not flexible)
- In choice of food item to use, use one you plan to make use of

Procedure :

1. Be sure to follow all of the Notes listed prior to the Procedure, follow safety guidelines, such as wearing goggles, and also have completed the Set-Up before continuing.
2. Whatever your choice of food item, it will be referred to simply as chocolate from this point on for reference purposes in these directions.
3. Using a microwave-safe baking dish place chocolate in the bottom (note it is best to use solid-type with no other items in it and that it is reasonably thick). It will cover as much of the area as possible, particularly spanning from one side to the other.
4. Place the dish in the microwave in the center, close the door and turn it on. The amount of time will depend on the food type, its thickness, and power setting chosen, but in virtually all cases will be less than 40s and normally only run in the neighborhood of 20s.
5. The key to the amount of time heating is not only that it is short in time, but will depend on you keeping a careful eye on the food item heating. At the first signs of melting in a couple of spots, it is time to stop it. Do not let it overheat, bubble or even liquefy to any appreciable amount.
6. Be sure not to touch the item or move it too much to disturb it since you must measure the distance between the spots.
7. The number of measurements will depend on the number of spots. Each set of two spots in sequence are to have their distance between them measured. This is referred to as a Trial in the Data section and denoted by the letter S.
8. So this means Trial 1 is for the distance between spots 1 & 2, Trial 2 is for the distance between spots 2 & 3, and so on.
9. The best way to measure is to place the ruler across the rim of the dish and read by looking directly down where the centimeter side is above the spots in question.
10. Realize you are not starting at '0' with each measurement. For example let's say spot 1 is at 4.3 cm while spot 2 is seen at 10.8 cm – how far are they apart. You need to find the difference of these numbers and this is S (for this example it is 6.5 cm).
11. To help in determining where the 'spot' is use the toothpicks and poke them into the center of the melted spot on the chocolate, if soft, and if not, merely touch down to the center-point of the spot in question when reading its value on the ruler. Note do not force it into the chocolate as this may affect the readings of not only this spot but others. – In essence the toothpicks are guide posts. – Use only if needed.
12. Once all the measurements are done, look at these values – they should be similar to each other if done correctly. Determine the average of these values (X_{ave}) this is the value of half of the wavelength of the microwave in question.
13. Now double the average value and this is the wavelength of the microwave wave for our equation ($\lambda = 2*X_{ave}$) [Be sure to convert this to meters]

14. With the wavelength determined, we can determine the speed of light from the wave equation since we have either read the frequency from the manufacturer on the microwave or are using the standard value provided in the table below.
15. How did your results turn out? Use the percent error formula to determine how close you came expressed as a percentage. Be sure to only use two significant figures as this is the extent of your measurements as well.
16. Be sure to properly clean up and reset the microwave. It is more fun to use food that is being used in a fun recipe. Enjoy! :)

Data :

Frequency of Microwave [f] : _____ (Hz)

Trial	S (cm to nearest 0.1) Distance between melting spots
1	
2	
Average	

Calculations :

Be sure to Use Your Slide Rule! :)

Average Value :

$$X_{ave} = \frac{\Sigma S}{n}$$

(S is the distance between points, n is the number of trials)

Wavelength :

$$\lambda = 2*X_{ave}$$

Wave Speed : (Speed is Wavelength times frequency)

$$v = \lambda*f$$
$$c = \lambda*f$$

Percent Error :

$$\%E = \frac{[\ Actual-Experimental\ ]}{Actual} * 100\%$$

Constants & Conversions :

$1M = 10^6$

$1G = 10^9$

$1m = 10^2$ cm

$c = 3.0 \times 10^8$ m/s (value to use here)

    $[c = 2.99792458 \times 10^8$ m/s $]$

$f = 2450$ MHz (most microwaves are this)

    [use this value if unable to find or read yours]

Conclusion :

How did your results turn out (percent-error wise)? How similar were the melting points in the measured distances between them?

Activity #6
Determining the Wavelength of Laser Light
Grade Level : High School
Math Level : Challenging

Determining the Wavelength of Light in a Laser

Waves, in general, are phenomena in nature that transport energy from one place to another. Sound waves are compression or longitudinal mechanical waves that are vibrations of the medium they travel through (solids, liquids, or gases). Electromagnetic Waves are not mechanical and hence do not need a medium to travel through. All electromagnetic forms, like light, travel at the speed of light which is approximately 3×10^8 m/s. [More description depth of Waves is found in the Exploring Waves with a Spring Activity #38]

All waves also exhibit other unique behaviors in the presence of matter, such as reflection, refraction, and diffraction. In this Activity, the behavior of a wave passing through a narrow opening, commonly called diffraction. Defined **Diffraction** is the bending of a wave around a barrier, like the edges of a barrier. This is a behavior that helped Christian Huygens win the day over Newton's idea of how to characterize light. Newton had proposed a model in which light was composed of individual particles. If it were as it passed through a narrow opening, the most that would happen is that the majority would continue in a straight line, while some would create a fainter and fainter spray pattern from the center.

What was found, though, when light was passed through a single narrow opening, and in a later experiment using a pair of slits, was that there was a central bright region, called **the primary maxima**, *but on either side of it* were **bright spots separated by dark gaps**. The first pair either side of the primary maxima are called the 1st maxima, the next are the 2nd maxima and so on in terms of names for them. Huygens' Wave model of light could explain this phenomenon through **constructive and destructive interference of the waves as it passes through the narrow openings**.

The best way to describe why this occurs is to use **interference of waves. Waves can interfere in a constructive or destructive manner**. In Constructive Interference, this is the addition of two or more waves since their amplitudes are in the same direction. These result in the bright spots. In the case of Destructive Interference, two or more waves have amplitudes in different directions hence subtract or diminish the outcome. This corresponds to the dark region between the bright spots.

Instead of using a single or double slit arrangement, this activity uses what is called a diffraction grating which can have several hundred lines per millimeter! You have to know the number of lines per millimeter to successfully do this activity since its number will affect the pattern produced. A good thing to do is to have two different gratings and try the activity twice to see the outcome.

When using a diffraction grating, the large number of slits results in the same sort of diffraction pattern (dependent on the number of slits) that occurs with one or two slits. The geometry of this situation allows one to find the wavelength of the light itself! This is because the distance traveled by one beam through one slit corresponds to a distance equal to a multiple of the wavelength of the light from an adjacent slit. They are in what is called phase with each other. These waves will add up on the wall constructively where they hit and produce the bright spot.

This Activity takes advantage of the diffraction behavior of waves and their resulting interference in order to determine from geometry and careful measurement the wavelengths of light, such as that from a laser.

A quick question must be asked and addressed which is : Is laser light any different than ordinary light? Laser light is actually a terms that is an acronym. It stands for **Light Amplification by Stimulated Emission of Radiation**. It is a means for emitting electromagnetic radiation (like any ordinary light does) with a little more focus as compared to ordinary light. In essence, it is concentrated light. First, laser light has waves that are in step with each other, this is called coherence. Light from a bulb is not this way and given off in a random manner, hence might be called incoherent light. This coherent light is composed of waves of identical frequency, phase, wavelength, and polarization. Though this may seem different, it is not. We are using the best source of light so that we have consistency in the frequency and wavelength plus it is a naturally narrow beam. Caution must be practiced however so do not look into it or even let it bounce off and strike you in the eyes from highly reflective surfaces.

Purpose : To use the property of diffraction of light and geometric optics analysis to determine the wavelength of a common laser.

Safety Notes : Have Parent permission and supervision for this activity. Do not shine laser light into anyone's eyes. Always exercise caution when using a laser.

Materials :

- Laser (HeNe, Green, or other),
- Meter Stick,
- Diffraction Gratings (of known slit width, marked lines/mm),
- Measuring Tape,
- Table,
- Tape,
- 2 Pieces of cardboard or 2 toilet paper towel tube,
- Roll of Paper or taped together sheets to act as screen which will have marks made on it (dull finish to paper is best),
- Slide Rule

Laser Information & Other Pre-Preparations :

1) Can use regular small lasers. The key is that it must be placed in a v-shaped box parallel to the table (see step 3).
2) Wavelengths of various common lasers that may be used : Red HeNe Laser : 633 nm, Green : 543.5 nm, Orange : 612 nm, and yellow : 594 nm (1 nm = 1×10^{-9} m)
3) For the v-shaped holder : Either use a thin piece of cardboard folded in an accordion fashion so that it creates a v-pocket for the laser OR split a cylindrical toilet paper towel tube vertically and use each half circle as sides so that looking down on it there is a channel that is v-shaped.
4) Create a 2nd v-shaped form which the diffraction grating can set in. Note that when activated the laser has to hit and go through the grating, so work with the cardboard pieces or tubes so that this happens. A simple solution is to place the laser v-shaped platform on a thin book so as to elevate it slightly.
5) Diffraction gratings can be found on line easily at low cost. Average ones are $2-$3 each and have 13,500 lines per inch (1 in. = 2.54 cm) – Note : Depending on cost, using 2 different gratings in 2 separate trials (i.e. different lines per mm) is a good idea for comparative purposes. Record this value (q) on your data table.
6) Place the Table adjacent to a wall long-wise. On the wall tape the paper roll to act as a screen across greater than the width of the table (over 1 m and up to 2 m in length). (Best to use dull finish paper).
7) Note : Test the system so that a pattern is seen and works well. There is no need for further adjustments once the pieces are all in place and operational.
8) **SAFETY at all times – Must have parent permission and supervision for doing this activity. Lasers are not to be looked at or directed into anyone's eyes. Also be cautionary with laser light bouncing off metallic or shiny surfaces.**

Procedure :

1) Read through and conduct the pre-preparations so that the laser goes through the diffraction grating and has various maxima on the screen. Be sure to have measured the distance from the diffraction grating to the central maxima (L).
2) Note : Do not measure with the laser on. The distance (L) is measured with the laser off.
3) It is best to mark where the central maxima is with a pencil on the paper screen and the central portion of the 2 left and 2 right maxima (you can do more if interested). Be careful in marking and do not look into the laser. Use of a dull finish paper is best so as to avoid reflection. Face only the paper when marking it and do not look back to the laser.
4) Turn off the laser and take down the paper screen for measurements.
5) Note : You are not taking measurements with the laser on. You only mark the paper where the maxima fall. Measurements are done after the laser is deactivated and the paper is taken down to be measured.
6) Once all markings are done fill in the data table from measurements taken. Though it is noted to the meter, realize that you are measuring with a tool that can measure to the 0.1 of a mm.
7) The Data your are recording here is the distance from the central point to each of the subsequent maxima (x).
8) Follow the directions in the calculation section to determine the wavelength of light for the laser and compute percent error.
9) Calculations :
10) Though the Slide Rule is a recommended tool, all of these calculations can be done with a regular or scientific calculator. Some scientific ones even have built-in averaging formulae. For those who like spreadsheets, the data can be typed in and the formulae then also be typed in its own cell where the formula references each of the measured variables in their respective cells, for example B1..BN has the measurements and values used in the equation while BN+1 has the formula for all of these variables (why not A? Simple – use it to label you variables)
11) Determine the distance between the lines on the diffraction grating (d). (Be sure to convert to m).
12) Determine the distance from the diffraction grating to each of the maxima (L_o) (which is the hypotenuse) using the Pythagorean Theorem (the other sides are L and a given x)
13) You can either determine the sine of the angle in question from your data are merely move ahead and use the wavelength formula (since the sine is now expressed as a ratio of the measured and calculated sides of the triangle – see diagram).
14) With a calculated wavelength (λ) compare it to the known values and compute percent error.
15) An alternative is to determine all of the wavelength values for the maxima used and then average the values as well.
16) If other diffraction gratings : Other Experiments : If you are using more than one diffraction grating , redo all the steps and have a new screen.

Photo for Set Up :

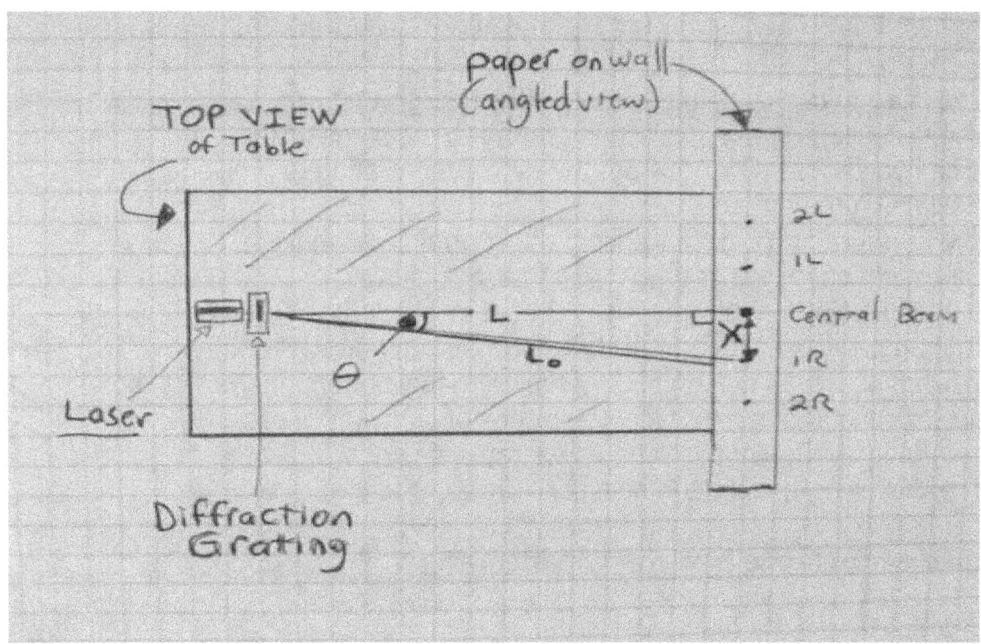

Data :

Number of lines per meter on the diffraction grating : _____ [q]
(Note – read the grating, conversion needed)

Spacing of Lines in Grating [d] : _____

Distance from Grating to Screen [L] : _____ m

n	x (m)
1 (L)	
1 (R)	
2 (L)	
2 (R)	

L : Left of center, R: right of center
Note that 'n' is 1 or 2 in the given case.

Calculations :

Be sure to use your Slide Rule!

1 m = 1000 mm
d : The spacing between lines on the diffraction grating (m)
L : The distance from the diffraction grating to the screen (m)
L_o : The hypotenuse distance from the diffraction grating to a given
   maxima on the screen (m)
x : The measure of the distance from the central initial maxima to the
   maxima in question (m)
n : Is the order of the maxima (n=1,2,et al as needed)
Θ : The angle as measured from the central maxima at which the next
   maxima occurs as determined by similar triangles from analysis of the
   constructive and destructive interference of light waves passing
   through the diffraction grating. (° or rad)
λ : The Wavelength of light in question (m)

Needed Formulae to Find the Wavelength :

$$d = \frac{1}{q}$$

$$L_o{}^2 = L^2 + x^2$$

$$Sin\ (\Theta) = \frac{x}{L_o}$$

$$n*\lambda = d*sin(\Theta)$$

Wavelength Formula (derived) :

$$\lambda = \frac{d}{n} * sin\ (\Theta) = \frac{d*x}{n*L_o}$$

Percent Error Formula :

$$\%E = \frac{[\text{Experimental Value-Accepted Value}]}{\text{Accepted Value}}*100\%$$

Average :

t : number of trials

$$\lambda_{ave} = \frac{\Sigma\lambda}{t}$$

Possible and Good Approximations Formulae in this Activity :

1) If Θ is used in radians, then $\Theta_R \sim \sin\Theta^\circ$
   To use – find Θ from

   $$\tan\Theta = \frac{x}{L}$$

   $$\Theta_R = \frac{2*\pi}{360^\circ}$$

2) $\sin\Theta \sim \tan\Theta$ for small angles
   This means for small angles, the ST scale on the slide rule can be used. What is being done here is we are using L instead of L_o and saying :

   $$\frac{x}{L_o} = \frac{x}{L}$$

Conclusion :

The best way to conduct the activity is to do it more than once and be
careful and certain of your numbers. Know that there is a certain level of
precision in your measurements, hence this affects the accuracy in your results.
The best measure is to determine percent error in your wavelength
determinations.

Activity #7
Chromatography Activity
Grade Level : Middle School
Middle School : Calculating

Though an Activity on Chromatography seems to be out of place in a book on Waves, it is not. Chromatography in the original sense deals with color – here we are dealing with inks that are created as a combination of color and are separating them. The idea of color is a vital part of waves, much like is seen in the spectrum of light as found by Newton, or as we often call the spectrum, ROY G BIV.

Brief comments on the color of an object. Next brief on how waves add. Finally subtraction of color and pigments and the inks here.

Chromatography comes from the Greek meaning 'to write in color' basically. It is the collective term that represents a set of laboratory techniques used in the separate of mixtures. The mixture is often dissolved in a fluid called the mobile phase (water is the material and is the solvent in our Activity), which carries it through a structure holding a material called the stationary phase. (In our Activity, the mobile phase is water while the stationary phase is the coffee filter). The various components or constituents of the mixture travel at different speeds, which, in turn, causes them to separate. The component separation is based on differential partitioning between these phases and it is the slight differences in a given material's partition coefficient that results in differential retention (i.e. spread or separation differences in the colors of the ink in our Activity).

The use of chromatography dates to Russian scientist Mikhail Tsvet in 1900. He used it with plant pigments and used it to separate chlorophyll, carotenes, and xanthophylls. They happen to have different colors (green, orange, and yellow respectively) this gave the technique its name. Greater efforts at the process expanded its uses in the 1930s and 1940s and beyond.

The idea of chromatography is used not just in Paper Chromatography, but there are others, such as : Gas Chromatography, Column Chromatography, Displacement Chromatography, Thin Film Chromatography, and several others. The type of chromatography depends on whether the technique comes from the chromatographic bed shape (column, planar, et al) or by the physical state of mobile phase (gas, liquid, et al).

Chromatography is used in Chemistry and many of its branches such as : Forensic Science, Pharmacology, Toxicology, Agriculture, as well as being used in Quality Assurance of a substance.

Our Activity will explore the components of water-soluble inks that can be found in some pens and/or markers. Not only will we separate the components of such an ink, but also measure the Retention Factor (which is the ratio of the displacement of the midpoint distance of a component to the distance traveled by the solvent).

Purpose : To compare soluble inks through paper chromatography and determine the Retention Factor for each.

Materials :

- Pens with Black Ink (the pens can be markers, regular ink pens, and if they are the only ones available they can have different colors – blue, purple, et al),
- Drinking Glasses (tall and clear are best – can be glass or plastic),
- Scissors,
- Pencils,
- Clear Tape,
- Coffee Filters,
- Water (from tap),
- Slide Rule

Procedure :

1. In terms of preparation – first find a set of pens / markers to make use of for the exercise. Note that not all inks are water soluble, so may not work. The Retention Factor of a pen that does not work is 0.
2. You can have as many glasses as there are pens if you want to run the experiment on each of the pens simultaneously. If there are not enough, simply use one glass and do each pen/marker separately (to avoid any cross-contamination).
3. Using a coffee filter (or two, if needed – particularly if you want to run the experiment more than once and perhaps average the results of the two trial runs of the same pen – a good idea) cut the coffee filter in strips that are tall enough so that each strip can be suspended by tape from a pencil across the top of the glass and hanging to just touch the bottom of the glass. Note : All strips should be of the same length and width. The width needs to be no more than about ½" (about 12 mm).
4. Trial with a Given Pen :
5. Use the Pen/Marker in question and draw a line at the same point near one end of the coffee filter strip. It should be about ½" to ¾" (12 mm to 18 mm) above the bottom so that water can be put into the cup (before the strip is) so that the water is below the line drawn on the filter.
6. Note the last direction to see that water is put into the cup first and only to a depth so that it gets the bottom of the coffee filter wet up to BUT below the ink line on the filter.
7. With the water in the cup and the coffee filter with a pen mark line on it, the coffee filter piece is taped at the other end around a pencil and the filter is lowered into the water (leaving the pen mark above the water line at the start) and the pencil goes across the rim of the glass to support the strip.
8. Note that each trial with each pen should have fresh water and in the same amount and done in the same way when it comes to drawing a line on the coffee filter.

9. Through the process of absorption, the solvent water works its way up the filter paper, passing the ink, and then drawing it along. Due to the fact that some of the components in the ink are more soluble in the solvent they will rise (and more rapidly) while others that are less soluble will adhere to the filter paper (and more slowly).
10. Once the water reaches the top take the sample piece out and lay it on plain white paper.
11. First count and name the number of unique colors you find in a given ink (some can be 1, others 2 or 3). Place the names of color on the Data Table in the order of discovery when going from the initial ink line to the top of the filter paper.
12. Next Use a ruler and pencil to measure first the Distance Traveled by the Water on the Paper and record this measure on the Data Table (to the nearest $1/10^{th}$ cm).
13. With each color band area, determine the midpoint of the color band, mark it on the white paper with a pencil and measure the distance from the initial ink line to the Midpoint of the Color band. Measured to the nearest $1/10^{th}$ cm.
14. Record this measurement on the Data Table in the space associated with the already named color in.
15. As noted earlier it is best to do each of the pens at least twice and to then average your measures from each of the results for a given pen.
16. Using the Data for each of the Pens calculate the Retention Factor in 2 ways with the Slide Rule. First when placing the Ratio on the scales, find the smallest whole number ratio that is the same as this ratio by looking along the scales. Second with the ratio find its decimal equivalent. Write down both of these answers.
17. Another thing that can be done with the Retention Factor is to see how much more one is than another. With the various Retention Factors, divide all of them by the smallest one in the list using the slide rule and find the multiplier another is as compared to the smallest value.

Data :

Labeled Pen (A, B, etc)	Color of Line 1 & Midpoint Distance Traveled (cm)	Color of Line 2 & Midpoint Distance Traveled (cm)	Color of Line 3 & Midpoint Distance Traveled (cm)	Distance Traveled by Water on Paper (cm)

Calculations :

Be sure to use your Slide Rule!

Retention Factor (R_f) :

$$R_f = \frac{\text{distance traveled to midpoint of color line A}}{\text{distance traveled by water on paper}}$$

Average :

$$X_{ave} = \frac{\Sigma x}{n}$$

(x = a given value in the set, n = the number of elements in the set, Σ means the sum of the items)

Conclusion :

The first question to consider is this : Are all the inks composed of the same set of materials (note that different materials will result in different ink patterns)? The next question is this : What of the Retention Factor – are they the same or different? Why do you think one color has a different Retention Factor as compared to another?

Activity #8
Rate of Photosynthesis in Elodea Activity
Grade Level : Middle School
Math Level : Calculating

The connection of Waves to Photosynthesis is straightforward – the use of sunlight to effect living organisms that use it to manufacture food – and in the process also create oxygen.

From the Science of Biology, in green plants there is a process that occurs, called Photosynthesis. Photosynthesis is the chemical process whereby a main organelle in a plant cell called the Chloroplasts which contains Chlorophyll (a green pigment in plant cells) that traps light energy and uses it in a multi-step chemical reaction to convert Carbon Dioxide and Water into Sugar (Glucose) and Oxygen.

Plants do this in order to produce and store food. The plants take the sugar along with various nutrients from the soil to make starch, fat, protein, vitamins, and other complex molecules that are needed to help the plant grow, repair itself, and live.

The interesting byproduct is Oxygen, which many life forms need.

Needless to say, photosynthesis is the first primary biological use of the Sun's energy which acts as the base of the food pyramids and webs in nature. Herbivores feed on these sources, and in turn these grazers are fed upon by carnivores in the food chain. Hence (nearly) all the life on the planet has a connection to the Sun merely for food. This serves to illustrate the importance of the Sun which also affects its non-living environment and causes the Earth's overall temperature, weather, and climate that sustains life.

The chemical reaction of Photosynthesis is :

$$Carbon\ Dioxide + Water + Light\ Energy \xrightarrow{Chlorophyll} Glucose + Oxygen$$

$$6CO_2 + 6H_2O + \textbf{Light Energy} \xrightarrow{Chlorophyll} C_6H_{12}O_6 + 6O_2$$

The Activities explored here take a common aquatic plant found in many household aquariums, the Elodea, and examines the rate of photosynthesis by counting the number of air bubbles that are given off in a measured amount of time while controlling the distance a given light source is as the variable under consideration that may affect the rate of photosynthesis in Elodea.

In the Activity, the distance of the light source will be considered. The intensity of a light source was found to fall off at a rate equal to an inverse-square of the distance for the light source in another Activity. Here, various distances are considered to see if therefore light intensity affects the rate at which photosynthesis occurs.

Purpose : To measure the rate of photosynthesis in Elodea Plants and the change in photosynthesis due to the change in distance a light source is from the plants.

Materials :

- Sprigs of Elodea (3),
- Test Tubes with stoppers (3)(see Note),
- Rack for Test Tubes (this is one option OR use the large glass-),
- Large Clear Glass that the test tubes fit in (this will act as a mini-aquarium),
- Distilled Water,
- Baking Soda ($NaHCO_3$),
- Measuring Teaspoons & Tablespoon set,
- Measuring Tape,
- Stopwatch or Timer,
- Flexible Lamp or Overhead Lamp with 40W or 60W bulb,
- Slide Rule

Note : For safety considerations do not place the plants close to the light source being used. Also do not touch any item, tool, hand, et al to the light. Also do not position yourself so that when observing the plants, their stems, and the air bubbles that you are facing and looking into the light. Do not look directly into the light at any time.

Note : For each study, one plant is NOT a good choice. For any measurement, 2 or 3 plants at a time would be a better choice, each in its own chamber. Record each separately, but then graph the average result from these plants for a given measurement.

Note : Keep in mind that if you are working alone that you have to time yourself carefully and monitor the plants carefully – watching for the air bubbles at the stems during each two minute frame and keeping count. It may be best to have a tally sheet of some sort for this.

Procedure :

1. Note : A good choice for each of the Activities described here is to follow directions in placing the piece of Elodea in the test tube and Then –
2. Put the test tubes under consideration in a large glass or set of glasses that are filled with water when placed in front of the light source (this will help disperse the light and if used with good positioning might allow for the air bubbles to be seen easily)
3. In the glass, the test tubes can be taped with clear tape to the far side away from the light source (this can also therefore act as the test tube rack if there is none available).
4. **Activity : Effect of Distance of Light Source on Photosynthesis Rate**

5. Use 2 or 3 sprigs of Elodea that meet two criteria : 1) They are about the same size and 2) They each fit into the test tube individually

6. Note : The number you choose is the number you will observe and record data on. You will be monitoring the number of air bubbles that occur over 2 minute intervals and keep tally for each piece of Elodea separately, so choose carefully.

7. For each piece of Elodea – remove several of the leaves from around the stem.

8. Cut the stem at an angle so that it is like an incline that meets the other side (creating an area of exposure where the air bubbles will form and be monitored at). Also lightly crush this cut end to break it up a bit.

9. Use distilled water and fill the test tube between ½ and ¾ full. The amount depends on when the piece of Elodea is place in it and whether the plant piece is covered entirely or not – it needs to be submerged in the water. Note : add the Elodea after the next step.

10. The test tube should start with about ½ to 1 full of 1/8 tsp of baking soda and then next add the water and shake to mix.

11. Now place the Elodea in the water Stem side (cut side) up.

12. Note that to have a control have the entire test tube, water, baking soda, and Elodea system created, but also it will not be in the light at all. Be sure to monitor this one for its air bubble count for comparison. This is the Control. It is set away from the light entirely (in a dark place – to examine only briefly shine a light on it to note if there are any bubbles and how many).

13. Be sure to clear away the first set of air bubbles and wait approximately 1 minute once all is set up.

14. Either Set the test tubes up in the test tube rack that will be in the light or use the large clear glass system noted earlier. Even with the test tube rack, one could disperse the light and create a buffer zone between the test tubes and the light by placing a clear glass of water between them and the light source. Be sure to count this distance in your measurements.

15. Place the lit lamp at 100 cm from the test tubes. Note : Distance will depend on the wattage of the bulb – the higher the wattage, the greater the distance so as to not overheat the test tubes nor the rack. Monitor carefully and do not touch the bulb. Do not use a high wattage bulb.

16. With the lamp on, start the timer and begin counting the number of air bubbles that form at the cut stem end for each of the Elodea pieces. Keep a tally on a small piece of paper.

17. At the end of 2 minutes quickly mark down the results for each of the plants monitored and

18. Begin counting again for the next 2 minute segment. Record these results.

19. Continue this process for a total of 10 minutes. Then turn off the lamp.

20. Calculate the average number of bubbles per each time frame (N) for the number of pieces used.

21. Graph the Average Number of Air Bubbles (N) (y-axis) vs. Time (T in 2 min. increments) (x-axis) for a given distance.

22. Draw a best fit line through the points and determine slope. Note : Time is not 1 minute intervals, it is 2, hence you need to do some division.

23. Redo the Activity only changing the distance of the light source from 100 cm to 50 cm and next to 40 cm for a 3rd time, and perhaps even a 4th and 5th trials of 20 cm and 10 cm, time permitting.

24. Keep in mind that each experiment has its own line for a given distance.
25. Note : One does not have to continually change the Elodea being used between trials, but it would be good to wait a couple of minutes or more for a time off with the light off when it is repositioned and to freshen the solution that the Elodea is in so that this remains constant from one trial to the next – otherwise follow the directions and redo with fresh Elodea.
26. A second calculation :
27. Total the number of air bubbles from a given distance trial (for example all the 100 cm values) for all time frames (T). From this number determine the average number of bubbles for a sprig of elodea at that distance.
28. This now creates the variable the Average Total Number of Air Bubbles (N_T) for a specific distance.
29. Graph now the Average Number of Air Bubbles (N_T) (y-axis) vs. Distance from the Light Source (D). (Distances will be the various trials conducted, such as 100 cm, 50 cm, et al).
30. Draw a best fit line. Then calculate slope.
31. If, however, it appears to be an inverse-function exponential, then try the following :
32. Graph the log(Average Number of Air Bubbles (NT)) (y-axis) vs. the log(Distance from the Light Source (D)) (x-axis).
33. Draw a best fit line and calculate slope. If it is an inverse-square relation, the slope should be -2.

Data :

For the Activity, the following data table can be used. Note the inclusion here of doing more than one plant at a time and averaging the results.

Note : It is a good idea to label each data table, such as with the Distance of the Light from the Plants, or the Color of the Cellophane being used.

Time Frame (mins)	Plant 1 : No. of Bubbles	Plant 2 : No. of Bubbles	Plant 3 : No. of Bubbles	Average No. of Bubbles
0 – 2				
2 – 4				
4 – 6				
6 – 8				
8 – 10				

Calculations :

Be sure to use your Slide Rule!

Slope of Graph :

$$m = \frac{\Delta Y}{\Delta X}$$

Average Values :

$$M_{ave} = \frac{\Sigma m}{n}$$

(Σm is the sum of all the values in question for an average to be determined from, n is the no. of items used to determine the average)

Conclusion :

Looking at the data tables and the graphs, how does the distance (hence light exposure (see Inverse-Square Law of Light Activity)) affect the rate of photosynthesis of the plants?

Alternatives : Explore types of light (fluorescent vs. incandescent OR particular colors of light as an independent variable (note distance has to be constant here).

Activity #9
Various Rate of Heating Determinations
Grade Level : Middle School
Math Level : Calculating

Rate of Heating

The connection of Waves and Heat is quite clear – Infrared heaters are quite common and waves are seen as the carriers of energy as it is. In many of the following Activities there is the suggestion of using light bulbs and lamps – electromagnetic wave sources of energy. We can use quite a variety of lamps and bulbs for our experiments if materials permit. Also with the opportunity some of these studies could employ a natural light source, the Sun.

Why is the study of heat so vital? The obvious answer is nature and weather itself! Land and Water heat at different rates and the angle of the Earth's tilt with respect to the Sun plus its rotation creates a very complex system that results in different rates of heating along with the fact that the incoming radiation is partially reflected into space before even entering the atmosphere (about 30%). Different materials absorb at different rates and the energy can be redistributed, reabsorbed and retransmitted in the infrared (concerns about global warming due to materials even in the air such as CO_2, et al). The amount of water vapor, dust, along with the angle the sun strikes the surface, and the texture, and composition of the terrain plus the wind speed will affect the surface temperature. All of these affect the atmosphere's energy content, hence affecting pressure and pressure differences is the driver of the winds along with other factors too. This drives the weather!

Closer to home, heat and its transfer is critical in immobile structures (homes, buildings, et al) and even in mobile structures (cars, trucks, engines, et al). In buildings it is the retention of heat, particularly at cold times that is of importance. Yet when it is warm outside, the retention of cold air and the cooling of the air is critical. The measurement, dissipation, and control of the heat is very important. Even with a computer and its cooling fans and other appliances, such as refrigerators, and those that use heat like toasters, ovens, and coffee pots, heat once again becomes a factor of importance. The understanding, control, and use of heat is essential to many mechanical and biological items, aspects of life, and life itself.

The study of heat is a branch of physics called thermodynamics . What of the definition of heat and the units of heat? In physics, heat is considered the energy transferred from one body to another due to thermal contact due to differences in temperature. Naturally heat transfers from a body of warmer temperature to one of cooler temperature. When bodies are at the same temperature they are said to be in thermal equilibrium. When they are not the process of heat (energy) transfer can take place.

The examination of heat looks not only at appliances and buildings, but also weather, the interior of the Earth, and other bodies in the universe such as the Sun and other stars. Heat is a measure of the quantity of energy of a system, hence there is no 'cold' per se, merely the absence of heat as compared to another body. Also the Heat or Energy can only transfer via one of 3 known means : Convection, Conduction, and Radiation.

Conduction is transfer of heat/energy from particle to particle in a material or between materials in contact with each other. Often conduction occurs in solids. Metals are good conductors while wood typically is not. Convection is the transfer of heat/energy due to the movement of the particles of the substance itself carrying heated particles, such as currents in fluid materials (which includes liquids and gases). Radiation is energy/heat transferred by electromagnetic waves.

The units of heat are : Joule, calorie, kilocalorie. In fact, the factor noted before is the conversion for calories and joules : 1 calorie is 4.186 joules. Note that these are not the food calories we consume, each of these is 1000 of the base unit calories, hence the food ones are often written with an upper case 'C' as Calories. We could then convert our food calories to kilojoules by multiplying by the conversion factor. In many European countries, cans and boxes of food do not note the Calories, but instead the kilojoules instead.

Heat is something which may be transferred from one body to another according to the 2nd law of thermodynamics – essentially it moves from a higher temperature to a region of lower temperature naturally (i.e. the noted natural process and via one of the aforementioned means or if it does go from a place of lower energy to one of higher energy, it requires the input of work).

The following Activity set explores the energy of heat and more specifically the rate of heating of different items. Most of us have encountered and/or read about this. It is classically illustrated with a pot of warmed water and in it are two spoons – one metal with its handle out of the water and the other wood with its handle out of the water. In a short time frame the metal spoon is far warmer than the wood one, hence we classify metal as a good conductor of heat, while wood is rather poor.

As with all Activities, have your parent's permission and help in doing these Activities. Always employ safety in doing the Activities.

The key to each of these Activities as with most Rate type of Activities is that there is an Independent Variable that you control (the x-variable) and the Dependent Variable (the y-variable). Here the Change in Temperature in each case is the Dependent Variable.

Activity I : Rate of Heating based on Color of Object

Purpose : To compare the rates of heating of objects of different colors

Methods :

A) Painted sealed bottles or cans with water in them placed in sunlight (colors : silver & black)

B) Color tiles, felt cloth, plastic of same 4 colors (black, blue, green, white). Placed in pan of crushed ice (acts as snow). Light on it is constant for all. Measure depth with time for each color.

Notes for All Activities involving Light or Heat Sources : Safety first. Do not touch, stare, or put things on lit bulbs. Try to control the environment as much as possible by using a small pen light as your source when setting up and have the experiments in darkened areas as needed (the activity will tell you when this is critical, but it is fairly obvious).

Activity I : Rate of Heating based on Color of Object

Materials :

- **Activity IA :**
- Tin cans – radiation cans best (black, white, silver),
- Light bulb (about 100 W or greater) and a lamp where the shade can be removed,
- Water,
- Stopwatch,
- 3 Thermometers,
- Measuring Cup,
- Graph paper,
- Slide Rule
- **Activity IB :**
- Rectangular pan (glass preferable),
- Ice shavings or snow (if available),
- Ruler,
- Pieces of Felt Cloth approx, 2" on a side, colors : black, white, green, blue, (note : all materials are of same type and size),
- Overhead Lamp or sunlight,
- Note : The overhead lamp needs to have a flexible head to point the light source at the pan of ice shavings and strike an area equally where the cloth patches will be placed,
- Stopwatch,
- Slide Rule

Note : Activity IA : if school lab radiation cans unavailable, then use home aluminum or tin cans which had food at one time (such as canned vegetable 16 oz cans), now are empty, cleaned out, and made smooth edged in the area where a can opener operated. Next paint the 2 cans, 1 silver & 1 black & 1 white
with these, be sure to also create lids for them, such as using cardboard and poke a hole in the middle so that the thermometer can go through.

Note : Activity IB : Ice shavings can come from ice one makes at home and carefully smashing it up. A blender is a good idea instead.
The use of sunlight is best in this Activity since it will provide consistent lighting on all the cloth pieces simultaneously, but if not possible, use one or more overhead lights that illuminate the ice surface with the cloth on it as equally as possible. Desk lamps with lamp shade cones to direct the light may be of use here (use appropriate equipment and do not place things on lights directly)

Note : When using lights – employ common sense safety. Do not look into the lights, do not touch the light bulbs, do not touch items to the light bulbs.

Procedure :

1) **Activity IA :**
2) Each of the cans (either lab equipment or they are your own – see note above) must have the same amount of water. It is best if they are about half full.
3) They are to sit on small but equal stacks of books so that an unshielded light source sits in the middle of them where all the cans are equidistant from the light source. Initially the light is off.
4) Let them sit in the dark for an hour with the thermometers in them and then record the initial temperature readings for each of the cans in your table under the time '0' spot.
5) At this point activate the light. You have chosen your time frame for measurement (5 minutes works well). – Note it is best to shield the thermometer from direct light during the experiment so devise some shielding from cardboard or paper.
6) Check and record the temperature every 5 minutes for at least 1 hour.
7) Turn off the light. Put equipment away.
8) Graph the results as line graphs with Temperature (T) as the y-axis and Time (t) as the x-axis.
9) For each of the cans data, draw a best fit line (should be linear) and determine slope.
10) Compare and contrast results of the cans.
11) **Activity IB :**
12) Spread out the ice shavings in the pan to a uniform depth.

13) Turn the light on and in the circle of light on the ice shavings place the patches of cloth in a circular manner, but separated from each other so as not to affect each other. Be sure they are in the same radius region of the light so that they are receiving the same amount of light.
14) Every 5 minutes, measure the depth of the cloth patch from the top surface to the nearest 0.1 cm and record these measurements.
15) Record these measures for one hour.
16) When done, clean up and put away equipment.
17) Graph the data as separate lines on a line graph where Depth (D) is on the y-axis and time (t) is on the x-axis.
18) Draw the best fit line and calculate slope for comparison.

Data :

Activity I : Comparative Rates of Heating :

Activity IA : Heating Rate of Water in Can

Time Unit should be around 5 minutes, but consistent

Can   Temperature ($^{\circ}C$)

	Time 1	Time 2	Time 3	Time 4	Time 5	Time 6	Time 7	Time 8
Silver								
Black								
White								

Data :

Activity IB : Melting Rate of Ice under Cloth of a given Color

Time Unit should be from 5 to 10 minutes, but consistent

Depth (cm)

Can	Time 1	Time 2	Time 3	Time 4	Time 5	Time 6	Time 7	Time 8
White								
Green								
Blue								
Black								

Calculations :

Be sure to use your Slide Rule!

Slope $= \dfrac{\Delta Y}{\Delta X}$

Conclusion :

What do your results show you when it comes to color and heating rates
 for materials?

Activity #II : Rate of Heating with Distance Activity

Materials :

- Desk lamp with 100 W bulb and no shade,
- Measuring Tape,
- Tape,
- Measuring Cup (size can be ½ or ¾ cup),
- Water,
- 3 to 5 clear Glasses or clear plastic Cups,
- As many Thermometers as glasses (minimally 3),
- Slide Rule

Note : Safety first, do not look into the light and do not touch the light nor place anything on the light such as the cups or the thermometers!

Note : In all cases, if using plastic glasses, be sure to have sufficient distance between the bulb and the cup so as to not have any damage due to heat.

Note : If 3 Thermometers, they will be placed in glasses with water at distances of 15 cm, 30 cm, and 45 cm

Note : If 4 Thermometers, they will be placed in glasses with water at distances of 10 cm, 20 cm, 30 cm, and 40 cm.

Procedure :

1) The best set up is done in a room with no other lights affecting the activity other than the lamp with the bulb in it, so in the initial set up use a small pen light as a source.
2) See the Notes above with regards to safety and glass placement with thermometers.
3) In setting up the glasses let the thermometers sit in the water for some time (1-2 hours) in water with the lights off.
4) Once activated measure the temperatures at regular time intervals (for example 5 or 10 minutes is good). Record the data for each of the distances in question.
5) Be sure to measure the initial temperature of the thermometers once the light is activated.
6) After 1 hour, stop measurements. Clean up and put away equipment.
7) From the table of data, Graph all of the results. Each distance measured will have its own unique line of Temperature (y-axis) vs. Time (x-axis).
8) For each draw a best fit line, particularly through the areas of linear heating.
9) Calculate Slope of the lines and compare them.
10) First calculate the change of temperature (Final Temp at a given distance – Initial Temp at a given distance)

11) Next, Graph the Change of Temperature (y-axis) vs. Distance (x-axis). Draw a best fit curve through this curve (it should look like an inverse-square relation).
12) Determine the log(change of temperature) and the log(distance) for each of the values.
13) Graph the log(change of temperature) on the y-axis and the log(distance) on the x-axis.
14) Draw a best fit line and determine slope. (If it is an inverse-square relation, the slope should be -2).

Data :

Time Unit : _____ (minutes) (approx 5 minutes)

Record Temperature (°C)

Time Unit	Distance 1	Distance 2	Distance 3	Distance 4
0				
1				
2				
N				

Calculations :

Be sure to use your Slide Rule!

Slope $= \dfrac{\Delta Y}{\Delta X}$

Conclusion :

What relation did your data uncover for distance and heating from a source? Why do you think that this may be the case (see Inverse-Square Law of Light Activity for ideas)

Activity #III : Rate of Heating Comparison for Land & Water Activity

Materials :

- Two small Baking Pans of same size and composition (glass is best),
- Sand or potting Soil (enough to fill one of the pans),
- Water,
- 2 Plastic sandwich or snack bags,
- 2 Thermometers,
- Desk Lamp with no shade and 100W bulb,
- Stopwatch or clock to act as timer,
- Slide Rule

Procedure :

1) Like all of the experiments involving light on the object, find a place that has darkness for set up and let only the controlled light of the experiment be the only one affecting the soil/sand and water areas with their thermometers. For set up use a small pen light as a source.
2) In one pan fill to about 2/3 full with water while in the other loosely place the soil or sand (no need to pack it down) to the same depth.
3) Let the pans of sand/soil and water sit in a dark room for an hour or so before the actual measurements.
4) Place the thermometers in the plastic bags being used and put into the sand/soil. In the case of the water, you may have to tape the end down (duct tape is best – enough of it can protect the thermometer from the direct light when used here).
5) Be sure to leave the tops of the thermometers and tops of the bags exposed so as to have easy access to the thermometers to take out to read and be able to replace them.
6) Activate the light so that the pans are equally illuminated (hence the same distance. Be sure the thermometers are placed in some sort of biased way so as to be affected.
7) After 2 minutes start the initial reading of the thermometers. This is at time zero.
8) Start measuring time (every 5 minutes is good). At which times you record the temperature readings on the thermometers.
9) Record temperatures for at nearly an hour.
10) Turn off the light, dump out or put away materials.
11) For each set of data, plot Temperature (y-axis) vs. Time (x-axis).
12) For each separately draw a best fit line and compute slope.
13) Compare the rates of heating for each to decide which heats faster, land or sea.

Data :

Time Unit : _____ (mins) (choose a time interval like 5 minutes)

Record Temperature ($^{\circ}C$)

Time Unit	Soil	Water
0		
1		
2		
N		

Calculations :

Be sure to use your Slide Rule!

$$\textbf{Slope} = \frac{\Delta Y}{\Delta X}$$

Conclusion :

Can you draw any basic conclusions about the heating rate of water versus soil/sand? What if the water were deeper?
As a possibility, try this experiment outdoors in sunlight.

Activity #IV : Rate of Heating with Color of Light Activity

Materials :

- 4 Light Bulbs (white, blue, red, green or yellow) (Note : all of the same wattage value),
- Lamp,
- Thermometer,
- Measuring Tape,
- Stopwatch,
- Slide Rule

Procedure :

1) In this activity, it is best to do each of the measurements on separate days, but under the same basic conditions with the exception of the color of the light bulb being used.
2) Create an order to the bulbs to be tested (for example : white, red, then blue).
3) Set up the lamp with the bulb in question to be used. Plug it in. Note that you should not stare into the bulb when active nor touch it when active.
4) With the bulb off, but some sort of back light on to allow view (a small pen light is sufficient) set up the thermometer at a standard distance that will be used in all trials. (for example : 20 cm)
5) Note, you may have to mark this distance somehow and be sure to set up the thermometer in the same manner each time for each of the successive trials.
6) Turn on the light and after 2 minutes write down the temperature on the thermometer. This will be the initial temperature.
7) Every 5 minutes for 1 hour take the reading of the temperature. If it levels off before then, then discontinue.
8) Do this same set of procedures for each of the bulbs in turn.
9) For each set of results, graph the results : Temperature (y-axis) vs. Time (x-axis).
10) Draw a best fit line for the linear and climbing portion of the graph. Determine the slope of each of these lines.
11) What conclusions can you reach concerning the rate of heating and the color of light?

Data :
Chosen Time Unit : _____ (mins) (choose 3 to 5 minutes)

Record Temperature ($^{\circ}$C)

Time Unit	White	Blue	Red	Green
0				
1				
2				
N				

Calculations :

Be sure to use your Slide Rule!

Slope $= \frac{\Delta Y}{\Delta X}$

Conclusion :

What effects does color of light have on heating?

Activity #10
The Effects of Radiant Heat on Multi-Layered Insulation Activity
Grade Level : High School
Math Level : Calculating

In the wintertime we are very inclined to keep warm and one of the ways to do this is to 'bundle up' as our parents and grandparents tell us. We create layers to keep in the heat.

Astronauts have multilayered outfits to protect them from the harshness of space. For example, the astronauts on the Moon had to be able to protect themselves against extremes only imagined on Earth. In a region of darkness, it can be down to -100°C and in and in direct sunlight it can be 100°C.

Satellites orbiting the Earth have to endure the extremes of space as well. Oftentimes there are layered gold-colored foil on the craft. Though it appears flimsy, it is very critical to the operation of the satellites and probes. Its primary function is to reduce heat loss by thermal radiation. They are made to be highly reflective so that much of the incoming light is bounced off and away from the spacecraft. The layers are not designed to deal with the two other forms of energy transmission, conduction & convection. The layers are called MLI (Multi-layer Insulation). There are often several layers and they can be very close to each other, so long as they are not in thermal contact. They are very thin (to keep down weight, since this is an important issue in launching the device – every pound lifted into space requires 3-4 lbs of fuel). They tend to be made of very thin plastic on the order of micrometers in thickness (about ¼ mil) and made of Mylar or Kapton on one side and having a thin metal layer on both sides. The metal is often silver or aluminum. The average insulation blanket has 40 layers (our activity only examines 3). To help separation the layers may be embossed or crinkled (so they only touch on a few points) or can be physically help separated by a thin cloth mesh (scrim). The outermost layers are thicker and are reinforced with fiberglass typically. They are made with sewing technology. Layers are made, cut, stacked and sewn at the edges. The seams and gaps are the places most responsible for heat leakage. These heat blankets also act as shields by being the first line of defense to small dust particle impacts. Note these layers need to be grounded too so as not to develop a charge and cause radio interference.

In our Activity, we examine the main property of thermal radiation affecting aluminum foil (acting as our MLI) and measuring the rate of heating with time.

Purpose : To measure the rate of temperature change when a radiant heat source illuminates layers of insulation at a distance.

Materials :

- Aluminum Foil,
- Scissors,
- Flexible Neck Desk Lamp with Bulb (no more than 60 W),
- Tape,
- Meter Stick,
- Protractor,
- Marker,
- Lab quality Thermometers (up to 4 is useful, but this can be less),
- Stacks of Books or two inverted Crates,
- Timer,
- Slide Rule

Set Up :

- Either use 2 crates or two equal stacks of books and place the meter stick atop them. The meter stick will be like a bridge between either the two crates or two stacks of books.
- Measure the height of the crates or create a stack of books of a comparable height to the average crate.
- The goal is to have about a 8" gap between the table and the meter stick.
- For 3 layers of insulation roll out 3 sheets of aluminum foil 7" wide on a side each and then stacked atop each other.
- Use the Protractor as a template and trace it atop the first aluminum foil piece using a marker. Rotate it to make a complete circle image.
- Across the diameter create two tabs 1" across and 1" from the main circle.
- Cut out the 3 pieces together to be identical to each other.
- When completed they are 3 circles about 6" in diameter with tabs across from each other.
- Position the flexible light so that it shines almost directly along the meter stick nearest one end (in fact see note below – it is best to slightly angle it away from you so that it will not shine in your eyes when looking at the thermometers).
- At 10 cm from the light place the first thermometer (tied by string to the meter stick).
- Every 3 cm from the initial thermometer at 10 cm now place in succession, a layer of insulation (using clear tape – this is what the tabs are for to tape it to the meter stick) and then the next thermometer. End with a thermometer after the 3rd (or last layer)
- The aluminum foil needs to be perpendicular to the meter stick and to the light shining on it.
- Note : You can conduct this exercise with no thermometer in front of the insulation and only consider those behind it depending on the number of thermometers you have.
- Note : Another alternative is to first conduct the experiment without insulation for one thermometer, to act as a control and then to conduct the experiment with the 3 thermometers (depends on the number you have) behind insulation in a second trial.

-
- In the case of all thermometers try to find a way to angle them (if tall enough they can lean on the table, if not use a book alongside them or another meter stick). Angle them in a manner so that you can see them straight on to read them at any time needed.
- Test the light and the system only briefly.

Note : Safety is very important, so you must have parental permission and supervision in this Activity. Do not touch the light, do not touch anything to the light, do not look directly into the light, and set up the Activity so that you are not even looking indirectly at the light (see following notes).

Note : It is best to point the light at a slight angle so that when you are on one side looking at the thermometers it is not shining in your eyes, even indirectly.

Note : It may be a good idea even to do one of two things concerning the light :
   A) Wear sunglasses to block some of the light or B) Briefly turn off the
   B) light to take measurements for as short a time as possible, then turn it back on (note this will affect results some).

Procedure :

1. First follow the set up procedure. Be sure to only test the light and do not leave it on.
2. Note, if you are doing two separate trials (one for the control and one for the insulated thermometers, follow the same basic process for both in recording measurements).
3. If you can set up the system with the control in place do the following as it is and only modify if it is done twice :
4. Record all the initial temperatures of your thermometers after being in the darkened room for several minutes.
5. Now Turn on the light.
6. Every 5 minutes (have the timer going), record the temperatures of each of the thermometers. Be sure to not look into the light and take safety precautions to not look at it, have it affect your sight of the thermometers, nor touch the light.
7. Be sure to have positioned the thermometers so that they can easily be read and have little to no handling and do the process quickly when taking a reading (to keep time of measurement consistent).
8. Once done with the readings, turn off the lamp and let the entire system cool for awhile and have parental help in disassembly.
9. For each of the thermometer's sets of data graph as individual lines (can be done on the same graph, but often best to do separately) as Temperature (Y-axis) vs. Time (X-axis).
10. Draw a best fit linear line and determine slope.
11. Compare the Rates of heating to each other.
12.

13. You can also create a graph where the y-axis is the Rate of Heating (the slopes you determined in the prior steps) vs. the number of layers of insulation (0, 1, 2, 3) on the x-axis.
14. Draw a best fit line for this graph and determine slope. (Is the slope positive or negative and why do you think it is this way ?)

Data :

Note : the Number next to the T denotes the number of layers between it and the lamp!

Time (min)	T0 (°)	T1 (°)	T2 (°)	T3 (°)
0				
5				
10				
15				
20				

Calculations :

Be sure to use your Slide Rule!

Slope from lines graphed (Temperature vs. Time) :

$$m = \frac{\Delta y}{\Delta x} = \frac{\Delta Temperature}{\Delta Time}$$

Note : this calculation is done for each of the T lines separately!

Conclusion :

How does the rate of the change of temperature in time vary with increasing number of layers of insulation? Is it possible to decrease it to zero? What would happen if there was too much insulation (consider the craft and the need for materials and its weight)?

Activity #11
Graphically investigating the Inverse-Square Law of Light
Grade Level : High School
Math Level : Challenging

Inverse-square Law and Light Activity –

The heart of all of science is to not just notice phenomena, but the connections that a given variable has with other related variables. One of the most universal relations is known as the **Inverse-Square Law**.

In Physics, as well as all of the sciences, any physical law that states that some given physical quantity or intensity is inversely proportional to the square of the distance from the source of that physical quantity is said to be an **inverse-square law**.

The Inverse-Square Law basically applies when a given force, energy, or other conserved quantity is radiated outward in a radial manner from a source so that the surface area of this radiated area is spherical ($4*\pi*r^2$) hence proportional to the square of the radius, the force, energy or other quantity must spread out over this area. so this quantity must too diminish in intensity in an inverse proportional manner with the distance squared. It turns out that there are a number of phenomena that follow this general relation.

This activity explores the illumination of light, which all of us notice naturally. When we are close to a light it is clearly much brighter than when we are at a distance from it. **(Note do not stare into active light sources as this can damage ones eyes)**.

One of the first noted and mathematically explained **inverse-square relations** was the invisible force of gravity itself noted by Sir Isaac Newton in 1687. In his work, the Principia, he not only outlines his concepts of motion but discusses mathematically the force of gravity. He notes that the force of attraction of gravity falls off with the inverse-square of the distance between the centers of masses, such as the force between the planets and the Sun.

For example, this means that if there are 2 equally massed planets, yet one is twice as far from the Sun as the other, it would have only ¼ the gravitational force acting on it as compared to the other ($1 / 2^2 = ¼$). For the same mass at 3x the distance the force of gravity acting on it as compared to the closest one is only 1/9 as strong.

If on the other hand, let's consider 2 equal masses and move one mass towards the other, the amount of force rapidly increases. At half the original distance ($1/(1/2)^2$), the force of gravitational attraction between them is 4 times what the original force was! Recall that a net force necessitates acceleration (Newton's 2nd Law) – all the more reason planets closer to the Sun move faster than those farther away!

Newton's Universal Law of Gravitation :

$$F = G*\frac{m_1*m_2}{d^2}$$

Considering a Force with all other factors other than distance taken out :

$$\mathbf{F} \sim \frac{1}{d^2}$$

What is important about this relation in terms of mathematical analysis is that the product of the inverse-square of the distance and a given quantity (here force) will be a constant. We will make use of this in the Activity in terms of Light Intensity and the inverse-square of the Distances we use.

Later in history from Newton's time, August Coulomb found in 1785 that the attractive and repulsive force between unlike and like charges behaved much like gravity with regards to distance. That is to say, the amount of electrostatic force varies as the inverse-square of the distance between the charged objects.

Notice how fundamental forces like Gravity and the Electrical Force (2 of the 4 primary forces of the Universe) act in this manner. Imagine the surprise, excitement and wonder of the scientists who uncovered such things in those times. Nature was measureable, it was knowable, it made sense and seemed to have a simplicity and harmony to it.

It turns out that the Intensity of Light (which includes all forms of Electromagnetic Radiation) has this same relation between intensity and distance from the radiating source! With distance, the intensity of light falls off as the square of the distance.

Our goal is to examine the Inverse-Square Law and its application to the Intensity of Light. In the Activity, we use a Solar Cell (photovoltaic cell) connected to a Multi-meter that is at a measured distance from a luminous source (an incandescent lamp with a light that is on). We measure the Current (I) for the Solar Cell at the given distance (d) when illuminated by a light source. (Note : We are not using Voltage, since it will remain fairly constant with distances that are similar to each other as in our experiment (test this for yourself)).

It turns out that there is a relation between the square of the current and the power of the light. How this relation comes about is this way : Power for an electrical system can be determined as voltage time current (P = V*I) and from Ohm's Law (V = I*R) we can substitute for V the terms I*R and obtain (P = I^2*R). The resistance (R) here is treated as internal and constant to the solar cell (photovoltaic) and factored out, so P $\sim$ I^2.

In the case of light sources, Illumination (B) is the number of lumens per meter squared ($B = \frac{L}{4*\pi*D^2}$)(the units of Illumination are lux). The Luminance (L) aka Luminous Flux is proportional to the Power of the Light Source ($L \sim P$) hence $L \sim I^2$. Though we measure light sources in terms of Power (Watts) they can be considered in terms of their luminance. In fact for a given wattage of a incandescent bulb, they average about 15 lumens per Watt while fluorescent bulbs (depending on power rating) range between 50-100 lumens per Watt.

Our Activity will place a simple incandescent light source at a distance from a photovoltaic cell connected to a multi-meter set up to measure the current (in milliamps) at varying distances from the light source. Both the Distance (D) and the current (I) are recorded. This data is graphed, mathematically recalculated, graphed again in two ways to determine the relation between light intensity (taken as current) and distance from the source to find the inverse-square law relation!

It is useful to examine this phenomena since it involves the idea of inverse -square relations, graphical and mathematical analysis, plus in the case of light, it is practical since when coupled with other interactions of light with matter (particularly reflection, absorption & re-emission, as well as refraction) it helps in understanding light and can be employed for light intensity needs for indoor as well as outdoor lighting considerations.

Purpose : To investigate through graphical and numerical analysis from measurements using a photovoltaic cell and a multi-meter the relation of light intensity and distance from the source.

Materials :

- Short Lamp with bulb (60 watt) (take off shade),
- Measuring tape (metric measures are best),
- Multi-meter ,
- Photovoltaic solar cell,
- Paper Towel tube or Toilet Roll tube cut to 5 cm cylinder,
- Sheet of dark Construction Paper,
- Graph paper,
- Long table (kitchen will do),
- Inverted Crate or Stack of books (to set multi-meter and solar cell on),
- Darkened Room,
- Small LED light to allow for writing, reading of information,
- Clear Tape,
- Slide Rule

Procedure :

1) Note : have the room as dark as possible, but have a small LED light to be able to do work as needed.
2) First cut the paper tube (paper towel tube or toilet roll tube) to a length of not more than 5 cm.
3) Attach to one end the dark construction paper and cut out the hole.
4) This tube with a shield will be placed in front of the solar cell so that only a constant area with some protection will act to only allow in the light along that path and keep out all other stray light sources.
5) With lights in the room on, set the lamp at the far end of a table.
6) Unfurl the measuring tape from the base of the lamp away from it across the table.
7) It is best to use an inverted crate (or stack of books if the crate is not available) to place the multi-meter and solar cell on.
8) The Solar Cell must be set up so as to be at the same height of the lamp and facing it (the tape can help hold it up) (books, magazines and the like can be used for either the lamp or the crate to establish a level environment so that the solar cell and light bulb are on the same level).
9) Set the solar cell so that it is the prescribed distance as noted in the table for data.
10) Note that the tube assembly created in steps 2,3,4 is placed in front of the solar cell.
11) Attach the multi-meter so that it will read current (I) measured in milliamps.
12) Turn on the bulb and turn out the other lights in the room.
13) Note : Do NOT stare directly into the light when taking measurements.

14) For each reading, always let the system sit and stabilize for a few seconds. The values may oscillate around a couple of numbers – take the one that seems to be the most frequent.
15) Once you are done with all of the measurements, turn on the room lights and turn out the lamp. Disconnect the multi-meter and solar cell.
16) Follow the remainder of the directions in the Calculations portion.

Data :

Measuring the Inverse-Square Law Indirectly
(through Multi-meter Current Readings & Power Calculations)

Distance (cm)	Current [I] (mA)	Current^2 [I^2]	$\dfrac{1}{D^2}$
10			
20			
30			
40			
50			
60			
80			
100			

Calculations :

Be sure to use a Slide Rule for all of your calculations!

1) Use the Slide Rule and find the squared value of the current (I^2) and place this in the appropriate column. (This uses readings from the D Scale and looking at the A Scale).
2) Note that Electrical Power is proportional to the square of the current in the circuit. ($P \sim I^2$).
3) First use the D Scale looking up the distances used in the Activity and find the corresponding value on the C1 Scale (its inverse). Be sure to watch the exponents here! Fill in this portion of the Data Table.
4) Create a Table of log(distance) and the log (current^2) values as read from the Slide Rule. (Read the data from the D Scale and find the log values on the L Scale).
5) Graph on the Y-axis Current^2 (which is related to Power which is related to Light Intensity) vs. Distance on the X-axis. This curve should approximate an inverse-square relation.
6) To test the idea of the relation in the Activity :
7) Calculate the Product of Current^2 and $\dfrac{1}{Distance^2}$ values. If done correctly, they should be approximately the same. Add these values up and divide by the number of data points to determine an Average value.

8) Graph I^2 (Y-axis) vs. $\frac{1}{D^2}$ (X-axis).

9) Draw a best fit line and determine slope for this graph. It should be approximately the same as the average of the product!

10) Graph on the Y-axis the log(current^2) and on the X-axis log(distance).

11) Through the log-log graph data, draw a best fit line and calculate the slope of this line (which should be near -2).

12) Your relation will reveal the connection between the variables. If it is at -2, this means that the light intensity (which is measured by current^2) is inversely related (hence the negative sign) to the square of the distance (light intensity $\sim \frac{1}{\text{distance}^2}$)

13) Though the Slide Rule is a recommended tool, all of the calculations can be done with a graphing scientific calculator or the use of a spreadsheet program. In these calculations you have to generate a table of data, graph it, and then find the slope and/or equation of the best fit line for the data. Other formula calculations can be done with these tools as well.

Formulae :

$P = V*I = I^2/R$
   (Note 'R' here is assumed to be constant and factored out –
   we are noting that $P \sim I^2$ and that $P \sim$ Light Intensity)

$$\textbf{Product} = \mathbf{I^2} * \frac{\mathbf{1}}{\mathbf{D^2}}$$

$$\textbf{Average} = \frac{\mathbf{\Sigma(products)}}{\textbf{number of products}}$$

slope $m = \frac{\Delta Y}{\Delta X}$

$$\mathbf{m} = \frac{\mathbf{\Delta(I^2)}}{\mathbf{\Delta(\frac{1}{D^2})}}$$

$$\mathbf{m} = \frac{\mathbf{\Delta log(I^2)}}{\mathbf{\Delta log(D)}}$$

Conclusion :

Examining the data and the graphs one should be able to find the inverse-square law relation of light intensity and distance. Be sure to take into account all sources of error and redo this as necessary.

Activity #12
Light Inverse-Square Law Basic Approximation Activity
Grade Level : High School
Math Level : Challenging

In nature there are several physical phenomenon that when examined exhibit an Inverse-square relation between the variables in question. The classical ones known are the Force due to Gravity (Gravitational Force) and Electrostatic Charge or the Force between similar or dissimilar charges aka Coulomb Force. Each of these acts with an Inverse-square of the Force with distance. That is to say, if we let the unit of Force be 1 at a standard Distance of 1 unit, then the Force will be ¼ or $(\frac{1}{2})^2$ at twice the Distance (2 x).

For more information on the idea of the Inverse-square Law read the Prelude to the Inverse-square Law Activity for Light where a multimeter is used to find this relation and the discussion of the idea of the Inverse-Square Law relation.

The following Activity employs only a Flashlight, some cardboard, a pair of scissors, a measuring tape, a ruler, and a wall to find the same relation.

This Activity contrasts with the other Light Inverse-Square Law Activity in that this one does not measure the light directly as we did in the other Activity using a Solar Cell and measuring the current that is produced by the incident light on the solar cell on a Multimeter. Here we are making the natural assumption that only a given amount of light is coming from the flashlight and hence as the area of light increases with increasing distance from the illuminated wall, the light is spread over a larger surface, hence a lower intensity.

In our Activity we will take measurements of a mostly covered flashlight (only allowing a spot of light to get through) when cast upon a wall at sequentially increasing distances. Since the flashlight has the same amount of light coming out at any time, with the increasing distance you should find that the size of the spot increases, hence the intensity of light would have to decrease (it is being spread over a larger area). We will use the changing size of the spot's area (assumed to be a square box measuring 's' on a side (though it will probably be due to a hole punch and therefore circular)) to act as the area that the light beam is covering. A calculated Intensity per Unit Area value will be generated. This variable and the Relative Distance (where the closest is considered 1 since we will divide all values by the initial distance) will have Log values taken of each and these log values will be graphed on a log-log graph where the determined slope will show the connection between Intensity per Unit Area and Relative Distance, which should be -2.

Purpose : To measure, graph, and mathematically determine the relation
between light intensity and distance through graphing.

Materials :

- Flashlight (standard type works best),
- Thick dark poster board Card large enough to completely cover the flashlight beam (typically about 10 cm x 10 cm), - An easy alternative is the cardboard back of a pad of paper as a resource,
- Scissors,
- Hole Punch,
- Measuring Tape,
- Ruler,
- Wall to project the beam onto,
- Plastic Crate or Stack of Books,
- Slide Rule

Procedure :

1) Set Up for the 'mask' that covers the flashlight :
2) Trace the cardboard piece so that it is a bit larger than the flashlight face and cut the disk out.
3) Using the hole punch, punch a hole in the middle of the disk (or as near the middle as possible).
4) Tape the 'mask' over the flashlight face.
5) Test the flashlight and see what it is like when the light is on. Try various distances from the wall. Note : You will see a central bright 'spot' which will be the item you measure in the Activity. Typically there are rings or a faint disk around the bright one – these are to be ignored.
6)
7) Setting up the flashlight to use in the Activity :
8) It is best to use a crate (or if none then use a box or easy to move stack of books) that you can tape the flashlight on top of where the light with the 'mask' on faces off the edge and points towards the wall to be used. – This set up is used so you can measure the distance to the wall and the size of the spot on the wall more easily. Also this adds to the consistency and stability in your use of the flashlight and the spot projected on the wall.
9) To complete the set up for the Activity, unfurl the tape measure from the wall (0) perpendicular to the wall for up to 2m from the wall and lock it here.
10) Since you will have a spot projected on the wall and it will be dimmer with distance from the wall, it is best to operate in as low a light as possible in the work environment. Have a night light or another smaller flashlight available so as to be able to take readings and write measurements during the Activity and then the room's lights can be off or at least dimmed.

11) In the matter of your Data Table distances (d) it is best to set up a chart of chosen distances before actually doing the Activity itself. Also it is best to use common multiples. For example one option could be : 30 cm, 60 cm, 90 cm, 120 cm, 150 cm. Another option could be : 25 cm, 50 cm, 75 cm, 100 cm, 125 cm. Make a choice and put these on the Data Table.

12)

13) The Activity :

14)

15) Place the flashlight set up (on the crate) so that it is at the shortest chosen distance on your data table. With all things ready to go, activate the flashlight.

16) Measure the diameter of the spot on the wall to the nearest centimeter in size. Place this measurement on the Data Table (s). Note : In your measurements of diameter, always be consistent in the manner how and where you make your measures. To help this, you might consider taking more than one measurement (say 2 or 3) and then average these values and place this value in the Data Table.

17) When done with the first distance, move the flashlight set up (light and crate) by sliding the system to the next distance and repeat the measurement of the spot diameter process.

18) Continue this process until you have completed the Data Table. It is best to have at least 4 data measurements.

19) Once done, dismantle the system. It is best to put things away so that there is no mess.

20)

21) Calculations :

22)

23) The first calculation does not need to be done, but it does make the process of data analysis easier. The first thing to do is to create the 'Relative Distance' (rd) by using the shortest distance and dividing it into all of the other distances. Note : If you have chosen your distances well, then all of your answers are simply whole number multiples. (It is good practice to use the Slide Rule, though these should be easily managed mentally).

24) The next thing to do is to square the values of the diameter of the spot. We are going to assume that the spot is a square with dimensions of 's' on a side, hence area is $A = s^2$. (For purists, you can treat it as a circle and the area of a circle with known diameter reading is : $A = \frac{\pi * s^2}{4}$). (Squaring is easy with the use of the A & B scales read in conjunction with the C & D scales).

25) The third calculation is to determine the 'Intensity per Unit Area' (IA). We will assume that the flashlight has 1000 units of luminosity available and that all of it is contained in the spot we measure. With each of the Areas we have calculated in the prior step, divide that figure into 1000 and write the answer on the Data Table (IA).

26) Next graph the light intensity per unit area (IA) on the Y-axis and the relative distance (rd) on the X-axis. Draw a curve through this. Note it should not be a straight line.

27) Next Generate a Table of the log values of the aforementioned points. Here we want log(IA) and log(rd) values using the Slide Rule and the D Scale with the L Scale.

28) Now graph these points log(IA) on the y-axis and the log(rd) on the x-axis. Draw a best fit line and determine the slope of this line. (the slope should be -2 ideally which indicates an inverse-square relation between the variables).

29) A further check of the Inverse-Square Relation can be found by doing the following :

30) First use the D Scale looking up the relative distances (rd) used in the Activity and find the corresponding value on the C1 Scale (its inverse). Be sure to keep track of the decimals here! Jot these down someplace.

31) Look up the square of these inverse values by looking them up on the D scale and finding its square on the A scale. Be sure to watch your decimal values. Use this information and fill in this portion of the Data Table.

32) Graph these data points where the Intensity per Unit Area values (IA) on the y-axis vs. the inverse-square relative distance values you have calculated ($\frac{1}{rd^2}$) on the x-axis. Draw a straight line through these data points and determine the slope of this line (should be the same value as IA computed in the first case with the shortest distance used).

33) Though the Slide Rule is a recommended tool, all of the calculations can be done with a graphing scientific calculator or the use of a spreadsheet program. In these calculations you have to generate a table of data, graph it, and then find the slope and/or equation of the best fit line for the data. Other formula calculations can be done with these tools as well.

Data :

Distance [d] (cm)	Relative Distance [rd]	Measured Size of Spot [s] (cm)	Area of Spot [s^2] (cm^2)	Intensity per unit Area [IA] ($\frac{1000}{s^2}$)

Table of Log-Log Values to Graph

Log (Relative Distance) Log (rd)	Log (Intensity per Unit Area) Log (IA)

Tabulated Data Table to Examine Inverse-Square Relation Validation :

Relative Distance [rd]	Inverse-square of Relative Distance ($\frac{1}{(rd)^2}$)	Intensity per Unit Area [IA]

Calculations :

Be sure to use a Slide Rule for all of your calculations!

Formulae :

Relative Distance (rd) :

$$rd = \frac{\text{measured distance- } d_x}{\text{smallest distance } d}$$

Area of Spot :

$$\text{Area of Spot} = s^2$$

Intensity per Unit Area :

$$IA = \frac{1000}{s^2}$$

Average :

$$\text{Average} = \frac{\Sigma(\text{products})}{\text{number of products}}$$

<u>Slope :</u>

$$\text{slope } m = \frac{\Delta Y}{\Delta X}$$

$$m = \frac{\Delta \log(IA)}{\Delta \log_{10}(rd)}$$

$$m = \frac{\Delta (IA)}{\Delta_{10}\left(\frac{1}{(rd)^2}\right)}$$

<u>Conclusion :</u>

What did your results show? What happened to the size of the spot of light with increasing distance? What does this mean for the Intensity of Light hitting the wall with increasing distance?

When you initially graphed the actual data, how did it appear – as a line, a parabola, or some other curve? What relation did the log-log plot show in terms of the relation of the variables?

Examining the data and the graphs one should be able to find the inverse-square law relation of light intensity and distance. Be sure to take into account all sources of error and redo this as necessary.

Activity #13
Inverse-square Law for Sound Activity
Grade Level : High School
Math Level : Challenging

The heart of all of science is to not just notice phenomena, but the connections that a given variable has with other related variables. One of the most universal relations is known as the **Inverse-Square Law**.

In Physics, as well as all of the sciences, any physical law that states that some given physical quantity or intensity is inversely proportional to the square of the distance from the source of that physical quantity is said to be an **inverse-square law**.

The Inverse-Square Law basically applies when a given force, energy, or other conserved quantity is radiated outward in a radial manner from a source so that the surface area of this radiated area is spherical ($4*\pi*r^2$) hence proportional to the square of the radius, the force, energy or other quantity must spread out over this area. so this quantity must too diminish in intensity in an inverse proportional manner with the distance squared. It turns out that there are a number of phenomena that follow this general relation. The more extensive writing on this idea is found in the first Activity on Inverse-square Law for Light.

This activity explores sound, which all of us notice naturally (in a frequency range of 20 Hz to 20,000 Hz). When we are close to a source of sound, it is clearly much more intense than when we are at a distance from it.

It turns out that the Intensity of Sound (which includes all wavelengths of it) has this same relation between intensity and distance from the Sound source! With distance, the intensity of Sound falls off as the square of the distance.

What this means is this : Let's say we have an agreed upon measurement of sound intensity (I) at a given distance (d). If we then double the distance and use our measuring device again, we will find for the same source of sound (now twice as far away) it will have an intensity of ¼ (which is $1/2^2$) of the original intensity.

Sound Waves are mechanical waves and like all waves possess energy. Being mechanical waves means that they require a medium to travel. We normally refer to sound travelling through the air, but it can travel through liquids and solids as well. In the latter case of solids and liquids we experience, measure, and see the effects of these waves as earthquakes. Sound waves are clearly much less intense in energy, but measurable none-the-less.

Most waves are measured by their energy transport as energy that is passing a point. We use the term Intensity to describe this. Since we are measuring energy per unit area and they are moving so it is also per unit time. But energy per unit time is Power (joules/sec) so sound intensity is really power per unit area and the unit of area is the meter-squared, so the basic unit is joules per meter-squared and recall that a joule/sec is a watt, so we can also say watts/meter-squared.

To measure sound intensity we use the unit decibels but and it does not use these units of watts per meter-squared, however. Decibel levels are taken relative to a standard, such as the level at which we can hear so is expressed as a ratio of the intensity in question, L_2, and the base intensity L_1 which is the threshold value. This means that decibels are actually unit-less numbers.

Decibel Level $= 10*\log\left(\frac{L_2}{L_1}\right)$

Notice that the equation is not linear, but instead logarithmic. This is because human hearing operates in this manner – we can hear rather low intensity sounds and quite intense sounds.

Because sound is a wave it has the characteristic of spreading in a inverse-square pattern (see those other Activities and some of the above to understand this better as needed) the formula will come out and be rearranged this way :

$$L_2 = L_1 - \left[\ 20*\log\left(\frac{d_1}{d_2}\right)\ \right]$$

Our goal is to examine the Inverse-Square Law and its application to the Intensity of Sound. We will make a sound and measure it at various distances with a device that measures decibels. We will use the above equation to see how close we come to an inverse-square law relation.

Purpose : To take sound measurement data in terms of distance and sound level intensity (decibel readings) and to employ a known mathematical relation to see whether or not it follows the expected inverse-square law relation.

Materials :

- Sound Measuring device – this can be one of the following : 1) Multimeter with a sound meter (dB), 2) stand alone sound meter, 3) microphone connected to a stereo system with a decibel sound meter,
- Graph Paper,
- Measuring Tape,
- The room being used is best if it is tiled or cement floors (to prevent sound dilution),
- Consistent sound item : Can be one of the following : 1) a toy that has a clicking sound, 2) tapping a metal pan lid with a spoon as consistently as possible, 3) snapping one's fingers as consistently as possible, 4) other?,
- Slide Rule

Procedure :

1. It is always best to read through a Procedure, have an idea for the plan, gather the needed materials, and then work the plan.
2. Depending on the arrangement of the room used (best to have tiled, cement, or wood floors) choose the longest path for the measuring tape to be placed on the floor. It is best to have the sound meter in one corner while the sound source will be moving from chosen point to point on the measured out path.
3. Note the previous direction where the points of distance between the sound meter and the sound source are already chosen. You can use the table provided or develop one of your own.
4. Realize one possible problem with the experiment – as distance increases between the sound meter and the sound source, it would be nearly impossible for one person to do this – so it may be best to have a partner who does one or the other.
5. Another important idea is to not only do one reading per distance. It is best to do at least 3. Not all of the readings will be used in calculations – instead the average value for a given distance is to be used.
6. Be sure to test the system to see if operating the way it should and adjust as needed. One of the important things to note is that with the sound meter on, there is probably a continuous background noise (hence it does not read zero). If the system allows a zeroing out, employ it – if not – recognize this separate to the data table as what the regular reading is. Note : It does not have to be subtracted out, since it is part of each reading it essentially washes itself out of consideration.

7. At each given distance [d] perform the sound creation and record the sound level measurement [L]. Do this at least 3 times per distance.

8. When done, continue with calculations – or perhaps redo the experiment to see if the results are consistent. Another alternative is to try a different sound source at the same distances – note that this has its own table.

9. Before performing calculations, it may be a good idea to graph the data as it is to see what sort of shape it creates. Put the Sound Level Reading on the y-axis and the Distance on the x-axis and draw a line through the curve – it should appear as a inverse-square law relation curve (see those Activities for more information).

10. Calculations :

11. First determine an average value for each of the sound level measurements. These are used in the calculations.

12. Note : So as to not have to deal with sign conventions in the needed formula provided : first note that [ ] are absolute value bars

13. and then choose d1 < d2 or conversely L2 < L1

14. (hence L2-L1 will be negative and regardless of the outcome of the variables and the operations on them inside the absolute value bars, it will be negative in magnitude due to the negation outside them).

15. Go through your table and do each of the needed calculations using your slide rule. In each calculation the first data point used is first on the list followed by the one that occurs next (at a greater distance).

16. Given the precision of your tools used, you may have to use no more than 2 significant figures.

Data :

Distance [d] (m)	Sound Level Reading [L] (dB)
0.5	
1.0	
1.5	
2.0	
2.5	
3.0	
3.5	
4.0	

Calculations :

Be sure to use a Slide Rule for all of your calculations!

Formulae :

Determining Average :

$$X_{ave} = \frac{\sum_{i=1}^{n} x_i}{n}$$

Formula to Evaluate for Inverse-Square Law relation in Sound :

$$L_2 = L_1 - [\ 20*log(\frac{d_1}{d_2})\ ]$$

Note : So as to not have to deal with sign conventions first note that [ ] are absolute value bars and then choose d1 < d2 or conversely L2 < L1 (hence L2-L1 will be negative and regardless of the outcome of the variables and the operations on them inside the absolute value bars, it will be negative in magnitude due to the negation outside them).

Conclusion :

Examining the data and the calculations, plus a visual look at the graph one should be able to find the inverse-square law relation of sound intensity and distance. Be sure to take into account all sources of error and redo this as necessary. One of the ways to know that this is on track, with each doubling of distance there should be a decrease in decibel level of 6.

Activity #14
The Index of Refraction of Glass and the Slide Rule
Grade Level : High School
Math Level : Calculating

This Activity considers when light waves pass through a material. This is the behavior called **Refraction**. Defined it is : **the change in direction of a wave as it crosses the boundary between two media in which the wave travels at different speeds.** This along with other optical properties when waves encounter matter and surfaces can be explored in Ch V The Story of Waves.

Description of the Activity :

In the Activity, a block of clear plastic is to be used and looked through on graph paper on edge. This type of material is common to science materials supplies and is very low cost. The other items used are pins, protractors, rulers, pencils, and slide rules.

Read the directions carefully once or twice through, practice it, and then do it.

The goal of this activity is to determine the index of refraction for one or more transparent materials by looking through them and measuring angles of incidence and refraction.

> **Purpose :** To determine the index of refraction of a common piece of glass or plastic from measurements of the angles of incidence and refraction as a beam of light passes through a piece of glass.

Materials :

- block of Acrylic Plastic (from science supply stores),
- 2 Stick Pins,
- Protractor,
- Cardboard (or Styrofoam or thick foam-core Poster Board),
- Ruler,
- Graph Paper (for graphs and drawings),
- Pencil,
- Slide Rule

Procedure :

1) Place the cardboard (or other) first on the table.
2) Cover this in Graph Paper (or Other Blank Paper).
3) Place the block of material so that it sits with respect to the graph paper lines. Note it does not have to match up, except along one horizontal and vertical edge would help.
4) Trace the block on the graph paper.
5) On the opposite side of the block away from you at the table stick two pins into the board so that a line through them as points forms a slightly acute angle to the block's side surface. Note : You might choose their points before putting them in and mark these spots. You will have to draw a line through these points at some point, so have this ready to go.
6) Use a ruler as a sight and look through the side opposite the pins where you are at the pins through the block of material. Use one eye (keeping the other closed) is best.
7) Move until the pins line up in your sight along the ruler line. Note : If done correctly it should form an angle that matches the angle of the pins to the side.
8) Draw a line along the ruler in the direction of the viewed pins.
9) Remove the block.
10) Draw with a ruler, the lines to the sides of the drawing of the block on the paper. Note : Do not draw through it!
11) Instead where these exterior lines contact the sides connect these points as a straight line with the ruler.
12) Note : If done correctly, the exterior lines are parallel to each other and the line between them will have a slightly different angle.
13) Take the Protractor and measure the acute angle on the far side of the block (the one with the pins) from the side to the exterior line.
14) Note : This value is NOT recorded! It is Not the angle to the Normal! To find the angle to the normal, determine the compliment of the angle measured. ($\Theta_i = 90° - \Theta_{measured}$) Record this value.
15) Note : The design of this experiment will have you measuring only acute angles at each interface.
16) Now measure the interior angle opposite you from the side to the angle made by the ray passing through the block (this is the acute angle that is opposite the angle measured in the exterior).
17) Remember to take this angles compliment! ($\Theta_r = 90° - \Theta_{measured}$) Record this value.
18) Perform these same measurements for the exit side (the side facing your) in a similar manner. For each measure the acute angle and take its compliment then record its value in the table.
19) For this data set, you can determine the Index of Refraction for the block twice and compare values. Recognize that you will be using the Sine (S) Scale for finding the sin of the angles.
20) Redo the experiment with fresh graph paper and choose different angles for the pins in each trial.

21) For each trial, determine the Index of Refraction.
22) Graph the sin (Θ_i) vs. sin (Θ_r). Draw a best fit line and determine the slope. What is the value of the slope and how does it compare to your determination of the Index of Refraction?

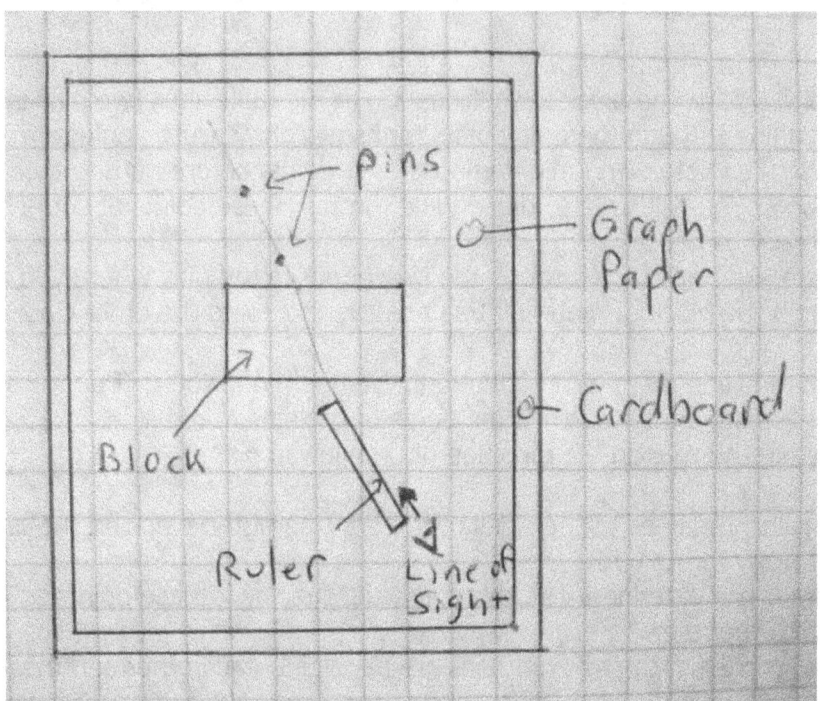

Data :

For each trial copy this data set.
Do this data set for at least 3 different angles.

Initial Side :

Angle of Incidence (Θ_i) : _____ ($^\circ$)

Angle of Refraction (Θ_r): _____ ($^\circ$)

Exit Side :

Angle of Incidence (Θ_i) : _____ ($^\circ$)

Angle of Refraction (Θ_r) : _____ ($^\circ$)

Calculations :

Be sure to use your Slide Rule for calculations!

Index of Refraction for Air $n_{air} = 1.0$

$$n_1 * \sin\Theta_i = n_2 * \sin\Theta_r$$

$$n_2 = \frac{n_1 * \sin\theta_i}{\sin\theta_r}$$

If you have the index of refraction of the material you are using :

$$\%E = \frac{[\text{Experimental Value-Accepted Value}]}{\text{Accepted Value}} * 100\%$$

Conclusion :

      The first thing to notice is the change of the angles as light passes through the medium. Did it become larger or smaller? Reread the Prelude for further insights into the probable outcome here. Next, compare the initial angle of incidence and the exit angle for a given trial. Also compare the internal angles of refraction in the same case. Are they similar or not? Finally, when you changed the angle and then both computed the index of refraction for the material and graphed it to find the slope, what value did it turn out to be? How close were the indices to each other and to the slope?

      If you have access to other transparent blocks of material do the same sort of exercise with them for comparison.

Activity #15
Converging Lenses Calculations with Diagram Method Activity
Grade Level : High School
Math Level : Calculating

This opening for this Activity is best found in Ch V Story of Waves and is about Optics, Lenses & Mirrors. The following Activity is about strictly Converging Lenses (though one can bring data for the other types of lenses and mirrors and do exactly what is outlined here)

The Activity :

We are going to draw pictures of what is happening and draw connecting lines according to the rules for lenses (in our case we are only going to look at converging lenses, but these rules apply to diverging lenses as well as concave and convex mirrors as well). They are called the Ray Diagram Method Process and Rules :

The Ray Diagram Method :

What is needed is a method to analyze the types of mirrors and lenses in a simple, mathematical way. That is where the aforementioned formulae come into play and in this Activity they will act as the accepted values. What we are going to do is be provided with information about lenses and mirrors – the type and with it two pieces of information, the focal length and either the object distance or the image distance. We can readily calculate the other, but we are going to employ a geometric construction method of drawing to find and measure the predicted outcome from the formulae.

In a Ray Diagram, the Lens or Mirror is placed on what is referred to as the Principal Axis (an imaginary line that the Object sits upright on (typically) and passing through the center of the lens or mirror). Along the axis is marked the Focal Point at the focal length distance and either the Radius of Curvature point (for mirrors) or Twice the focal distance (for lenses).

The Rays are the method of treating all electromagnetic waves of light as straight lines that project like a ray (an arrow tipped line segment). The principal rays are used only. When and where they cross will act as points of interest and where the solution lies. For example, say the focal point length and the object distance are given for a converging lens. Next, let's have the object distance fall between the focal length and twice the focal length. From the list above, we predict that the image should be larger, inverted, and more than twice the focal length distance. If we construct the drawing correctly, the 3 principal rays we use will intersect there. If we draw the drawing to some sort of scale, say each square equals one centimeter, then we can accurately measure what the image distance should be and our calculations should validate this.

In the following description, look to the images below them for help in understanding the process of their construction.

Rules for Drawing Rays for Lenses :

Ray	From Object to Lens	From Converging Lens to Image	From Diverging Lens to Image
1	Parallel to principal axis	Passes through focal point f	Directed away from focal point f
2	To the center of the lens	From the center of the lens	From the center of the lens
3	Passes through focal point f	Parallel to principal axis	Parallel to principal axis

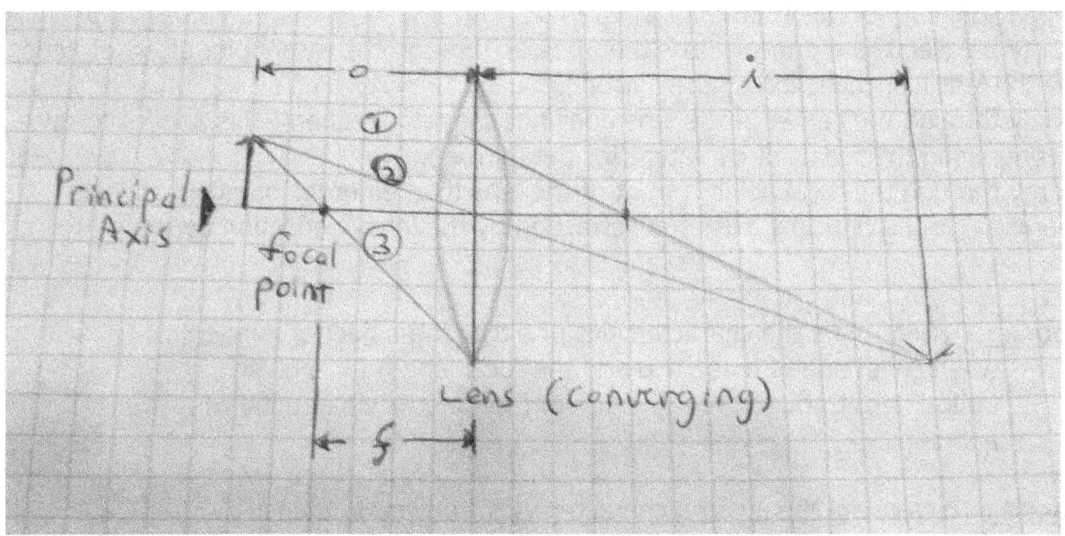

Rules for Drawing Rays for Mirrors :

Ray	Line drawn from Object to the Mirror	Line drawn from the Mirror to Image
1	Parallel to principal axis	Through the focal point f
2	Through focal point f	Parallel to principal axis
3	Through radius of curvature	Back along itself through the radius of curvature

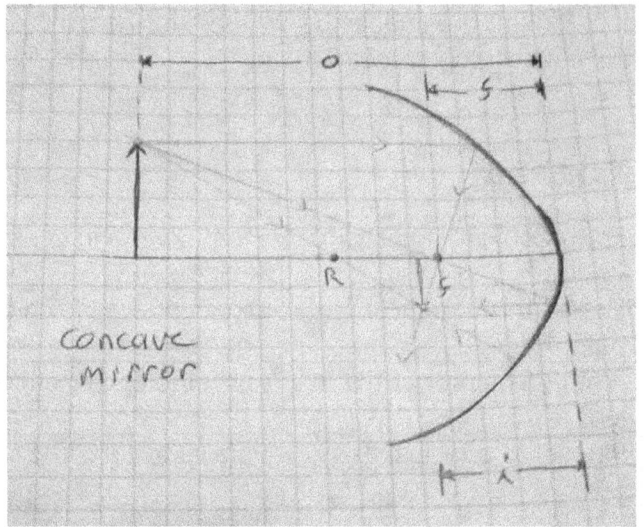

Here is a summary of these rules for diagrams :
1) A ray parallel to the principal axis then passes through the focal point after refraction by the lens (or reflection by the mirror).
2) A ray through the center of the lens does not change direction (this applies to rays acting along the radius of curvature for the mirror).
3) A ray through the focal point in front of the lens then emerges parallel (reflects parallel) to the principal axis after refraction by the lens (reflection by the mirror).

Purpose : To determine the characteristics of either the object or image when given one of these two and the focal length for a given optical item through the ray diagram (geometric construction) approach.

Purpose : To calculate the missing characteristics (object or image distance, object or image height, magnification) from given data and using the thin lens formula and magnification formula.

Materials :

- Graph Paper,
- Ruler,
- Table of Information to Use (Data),
- Slide Rule

Procedure :

1) For each of the object's below first read the prelude and predict the outcome of the missing information (object distance or image distance). Answer what type of image forms (real, virtual, inverted, upright), whether it is magnified or not.
2) The key is to draw 3 rays as described in the Ray Diagram method.
3) Draw your lens (or mirror – if you bring your own data) on a principal axis system and choose a scale for representation.
4) The first ray comes from the object or image parallel to the principal axis and then through the focal point.
5) The second ray travels through the center of the lens is a lens case.
6) A third ray goes through the focal point and either bounces off the mirror or goes through the lens and comes out parallel to the principal axis.
7) Next use the slide rule and calculate the actual answers to use as guides in your research for the answer.
8) For each, construct a Ray Diagram following the guidelines in the Prelude so as to be able to measure the answer and compare this to your calculated value.
9) To 'calculate' you need to only literally measure the distances of the object and the image (whichever is unknown in a given case).
10) Now use the Slide Rule to check the work. This can be done two different ways :
11) In the first way simple use the C and C1 scales to invert values as needed and then add and subtract them as you would decimal numbers (be sure to watch the decimal place on these) – In this first method you may need to read as many decimal places as possible for the best answer – be sure to use the next method as this will show where there can be some error.
12) The next way is to first algebraically manipulate the formula and solve the unknown quantity (below it is solved for 'f', though all other questions are for 'o' or 'i' – hint : both of these will look similar and look comparable to the focal point equation).
13) Determine a percent error. If low, then it is okay, if rather high, then return to both the calculation and the ray diagram and try again.
14) Note : The radius formula is for mirrors and not lenses.
15) Note : If you want, also calculate the object's or image's height and magnification from the formulae below.

Data :

Copy the Table below, Draw each situation, Measure your outcome,
& Compare your measured values against calculated values.

The Scale depends on the graph paper and the problem.
Sometimes it can be 1:1 (1 square = 1 cm), others it may be 1:2 or whatever is needed so that it fits on the paper, but keep it in mind since measurements are made with it!

Note : All Objects (or Images) have the same Height of 5 cm.
This is only the absolute value of the height and the sign will come from
the rules for lenses and mirrors as well as the formulae.

Object	Type	o	i	F
Lens	Converging	15		7.5
Lens	Converging	6.0		4.0
Lens	Converging	5.50		3.30
Lens	Converging	9.0		6.0
Lens	Converging	2.2		3.3

Calculations :

Be sure to use your Slide Rule!

i = image distance
o = object distance
f = focal length
m = magnification
R = radius of curvature (mirrors)

Thin Lens Equation : (Note : May need to manipulate equation first)

$$\frac{1}{f} = \frac{1}{o} + \frac{1}{i}$$

$$R = 2*f$$

Magnification equation :

$$M = -\frac{i}{o} = \frac{h_i}{h_o}$$

$$\%E = \frac{[\text{Experimental Value-Accepted Value}]}{\text{Accepted Value}} * 100\%$$

Example manipulated formula :

$$f = \frac{i*o}{i+o}$$

168

Conclusion :

From the determination of percent error (we assumed that the thin lens equation was the accepted values and the geometric construct measurements were the experimental values) we should see a very small margin of error. If there are large error, go back and try the process again for that item.

You can construct your own hypothetical situations (thought up or read about from books or the internet and can include all the other varieties of lenses and mirrors noted in the prelude) and try these.

Note that this is called the thin lens equation. Clearly there must be ones for thick lens equations as well. This also affects our outcome since this is an assumption that we have made for them.

Activity #16
Lens Considerations via Optic Bench & SR Calculations Activity
Grade Level : High School
Math Level : Calculating

In this Activity it is best to read the Ch V Story of Waves either in its entirety or at least the sections on Lenses, Mirrors, and other Optics topics. Here one needs to either buy a specialized tool, known as an optical bench (quite pricey) or use the materials list suggested and the set up described to have your own version of it and the only major purchase are the lenses you might consider to explore in using.

The goal of this lab is to use common tools to investigate the object distance and image distance produced by a converging lens and use this information to uncover from the thin lens formula the focal length of the lens itself.

Purpose : To determine the focal length of a converging (double convex) lens through object and image distance measurements.

Materials :

- **Option 1:** Optical Bench kit (from Science Supply companies),
- Other possible supplies if not in kit : batteries (usually AA or AAA, battery pack holder, et al),
- Slide Rule
-
- **Option 2:** (Your own Optical Bench made from list below) :
- Table,
- Meter Stick,
- Ruler,
- Marker,
- Small rectangular Styrofoam Blocks from hobby/craft store (often for plants) – these will be used to hold the lenses – several (good to have back up),
- Small collimated LED flashlight (one of the most costly items in this list),
- Convex Lenses of known focal lengths (>=2 of them),
- Slide Rule

Note : Always with all optical equipment and lights act in a safe manner – have parental permission and supervision.

Note : In the Optical Bench kit : There needs to be : Light Source (small LED lamp is best), lens holder, Meter Stick (best bench type), meter stick supporters, small screen for image projection

Note : It may be possible to buy the light source from a standard Optical Bench kit, so this may be less in cost than the collimated LED flashlight.

Set Up for Personally Constructed Optical Bench – Option 2 Materials :

- Use the materials so that the Styrofoam blocks act as lens holders.
- Be sure to draw a line across the blocks where the lenses are at so that they can be measured – the line points to the meter stick on the table they sit along.
- Use a meter stick on a table as the guide.
- This process of your own system will take some effort and improvisation at finding the right materials, the right alignment and height.

Procedure (for Optical Bench) :

1) Set up the meter stick optical bench according to directions.
2) Select a double convex (converging) lens and place it in the lens holder.
3) Record its known focal length for comparison.
4) Set up and test the light. Note it is best to use an LED type for this exercise. (May need batteries and battery holder pack then)
5) Place the image screen on the meter stick optical bench.
6) Choose an initial distance for the object (the light) that is greater than twice the focal length. Choose a number that is easy for recording and calculating purposes This is the object distance (o)(ex. If f = 6 cm, pick o = 20 cm).
7) Dim the room's lights and activate the light on the bench.
8) Move the screen to produce the expected image from the summary table of image characteristics (see Ray Diagram Activity).
9) Once focused, record the Image distance (i)
10) Note that for both the Object Distance and Image Distance that it is not merely the number on the meter stick since the lens is at a number as well. Distance is determined by the difference of these values.
11) Do this 3 times minimally. Have at least 2 trials at >2x the focal length and 1x between 1 & 2x the focal length.
12) Calculate the focal length from each of the trials.
13) Compare and compute percent error from the known value for the focal length of the lens.

Procedure for Option 2 – Personally constructed Optical Bench :

- Same as the Option 1 procedure – realize you have made the same type of system (so looking on-line is a good idea).

Data :

For each lens used :

Known focal length : _____ (cm or mm)

Trial	o (cm)	i (cm)
1		
2		
3		

Calculations :

Be sure to use your Slide Rule!

i = image distance
o = object distance
f = focal length
m = magnification

Thin Lens Equation : (Note may need to manipulate it first)

$$\frac{1}{f} = \frac{1}{o} + \frac{1}{i}$$

$$\%E = \frac{[\text{Experimental Value-Accepted Value}]}{\text{Accepted Value}} * 100\%$$

Conclusion :

Examine your calculated focal length value and compare it to the known value of the lens. How close did your experimental process and measurements come?

Activity # 17
Estimating the Size of the Moon
Grade Level : Middle School
Math Level : Calculating

Size of the Moon Activity –

On a given full Moon night have you thought about how far away the Moon is, how big the Moon is? This is the Activity to use basic math to use its distance and a simple observational trick to find what the diameter of the Moon is. The interesting part (beyond the math) is the fact we are merely observing the Moon and taking a few measurements as well as using extremely common and readily available items.

In our Activity, we will use a Proportion where we take the unknown diameter of the moon to its distance from the Earth as a ratio and set this equal to the known diameter of a punched out hole in an index card over a measured distance along a meter stick that the hole is from us when observing the Moon so that it just fits the hole exactly.

We will only use 2 significant figures, though a Slide Rule can go to 3 for this exercise. Those who want to take it further, use the more precise data.

Materials :

- Hole Punch,
- Clear Tape,
- Meter Stick,
- Index Card,
- Scissors,
- Full Moon,
- Distance to Moon : **2.4×10^5 mi (3.8×10^5 km)**,
- Paper, writing tool (to record measurements),
- Slide Rule

More Precise Data :

Earth-Moon Distance :
2.39×10^5 mi (3.84×10^5 km)

Data for Comparison (the Diameter of the Moon) :

Note : Be sure to go past this section until after you have taken your measurements and calculated the estimated diameter of the Moon from your data!
2 significant figure value : 2.2×10^3 mi (3.5×10^3 km)
3 significant figure value : 2.16×10^3 mi (3.47×10^3 km)

Procedure :

1) Use a ruler and measure the width of the meter stick (yard stick) you are going to use and write it down on scrap paper for use.
2) Knowing the width of the meter stick, measure this amount twice from the edge of a 3x5 card so as to mark the 3" side with two marks.
3) Draw these as two parallel straight lines across the card. (see 1st photo)
4) All of the remainder of the card beyond the second mark line is to be cut away and discarded.
5) In the middle one of the drawn strips, place a piece of clear tape that is about 1 cm square in size.
6) In the center of the tape, punch out a hole.
7) With the ruler measure the diameter of the hole (H) and record it in the data table.
8) Note : In your measurements you only have to be consistent. That is both the hole diameter and its distance must be in the same units. You can use the metric system or the English system.
9) Cut along the line between the two strips so that only enough of the middle remains on the top of the meter stick and the remainder acts as flaps that wrap around the meter stick.
10) Place the meter stick flat on a table. On top of it place the card so that the hole is centered on the stick.
11) Fold along the line so that the hole is standing up perpendicular to the meter stick.
12) Wrap the other strip in contact with the meter stick around the meter stick and tape it so that it becomes a firm wrap that can be slid along the stick. Cut away excess so that it is a firm sleeve for the meter stick. (see 2nd photo)
13) If correctly done you have made a stand up hole to view through from the end of the meter stick which can be slid along the stick as needed. (see 3rd photo)
14) When using it be sure to keep the hole portion perpendicular to the surface of the meter stick.
15) On a clear night with the full Moon out go outside and point the stick at the Moon so that you have to look along the stick and through the hole in the card at the Moon.
16) Be sure to use one eye looking along the stick and through the hole at the Moon.

17) Move the hole so that it just matches the size of the full Moon. Be careful not to move this once in place since its distance (D) is needed in the data table. Record this value.

18) Note : Be sure to measure it correctly. The point that the hole is over is the distance from the end that is zero (if looking from that end).

19) With data in place calculate the diameter of the Moon using the formula provided and using your Slide Rule.

20) Note the ease of using the Slide Rule since the formula is a proportion!

21) Note : In using the Moon Diameter Formula, pick which units you want your answer to be in by choosing the distance to the Moon measure as either kilometers or miles. Whichever is chosen is how the answer will come out.

22) Check to see how accurate your results are by using the percent error formula (again using the Slide Rule).

23) Note : Though the Slide Rule is a recommended tool, all of these calculations can be done with a regular or scientific calculator. Some scientific ones even have built-in averaging formulae. For those who like spreadsheets, the data can be typed in and the formulae then also be typed in its own cell where the formula references each of the measured variables in their respective cells.

Photos

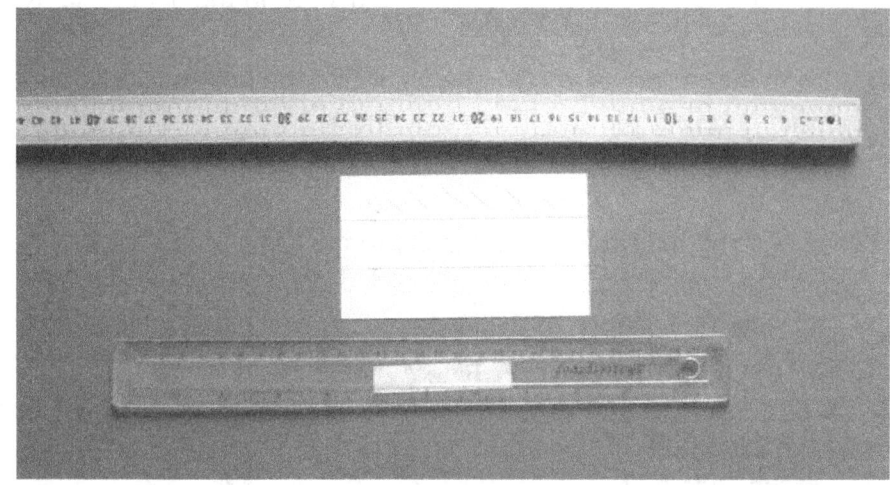

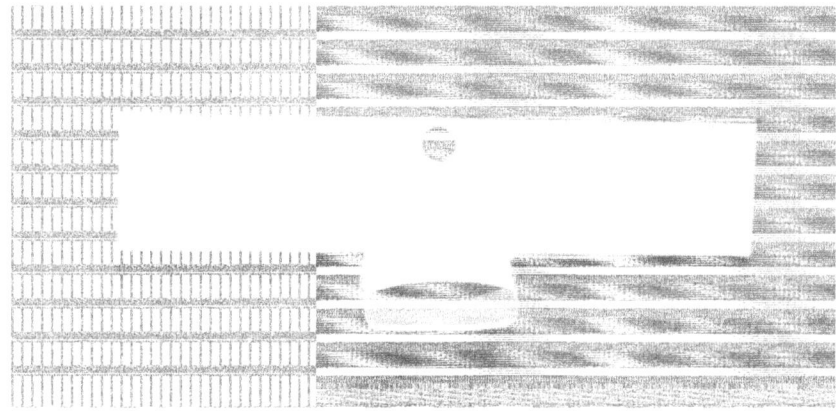

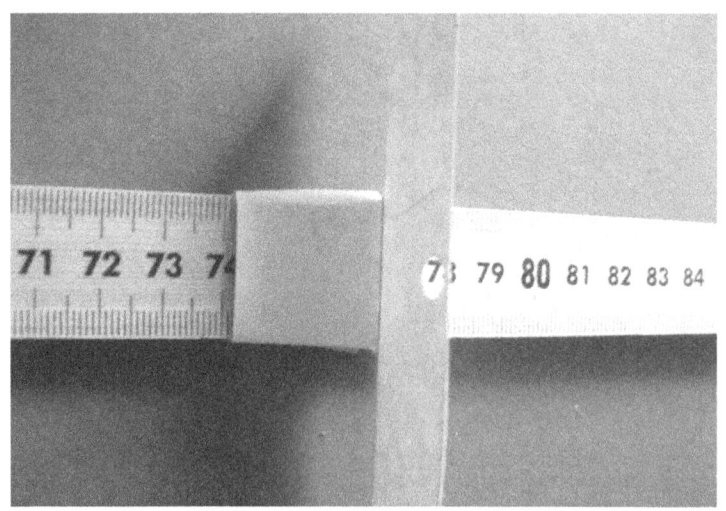

Formulae :

Be sure to use your Slide Rule!

Moon Diameter Proportion Formula :

$$\frac{H}{D} = \frac{\text{Unknown Moon Diameter}}{\text{Known Moon Distance}}$$

(H : hole diameter, D : distance of hole to eye on meter stick)

Percent Error Formula :

$$\%E = 100\% * \frac{[\text{Accepted Moon Diameter - Calculated Moon Diameter}]}{\text{Actual Moon Diameter}}$$

This is an alternative or an add-on to the Activity noted in the Quest for a New Tool (see that article for more information). Some of typical measures for this Activity are these : 0.5 to 0.8 cm for the hole diameter and the typical distance range for how far the hole needs to be is > 55 cm and < 95 cm. This is because the distance for this ratio of Earth-Moon distance to the Moon diameter is about 111 times.

This can be seen as asking the question, 'how far would a dime seen face on need to be in order to cover the full Moon?' The dime is nearly 2 cm (around 1.8 cm) so it would have to be nearly 200 cm (2 m) distant.

Activity #18
The Size of the Moon using a Telescope Activity
Grade Level : High School
Math Level : Calculating

In this Activity, you need to have a Telescope (Reflector or Refractive Type) (For information on Telescopes, see Ch V on the Story of Waves) and some knowledge of its operation. The basic goal is to go out during a time when the Moon is out and set up the scope to look at it. It will not matter the size of the eyepiece, though it is recommended to be a low power variety. Often this is somewhere between a 18mm or 25mm size.

Once the Moon is in sight, the basic goal is to align the scope so that one edge of the Moon just matches the edge of the field of view. You can choose to leave the Moon in sight and use the 'right' edge or have the Moon just out of view and use the 'left' edge. With your telescope on the object, as the Earth rotates, the Moon will then move out of view or into view (depending on your alignment choice).

Now use a stopwatch and time this event. How long does the Moon take to either move out of view or just into view?

Once we have time, we will then make some simple assumptions. We will assume that the Earth rotates at a constant speed and does so in exactly one day. With this idea, we then are using the rate of $15°$ per hour for the rate of rotation (Since $15° \times 24 = 360°$).

We will also assume that during this transit time, the Moon is not moving relative to the Earth (since it is, this will become one of the factors affecting our ability to measure its actual diameter).

The reason for having time is part of the goal of the Activity. We will use the time measure to determine the amount of angular movement of the Earth across the diameter of this object and translate time to an angle then to a diameter for that object (the Moon in this case)!

Taking the measured time value and converting it into radians (see Procedure) we can then readily find the approximate diameter of the Moon (assuming we use the known distance to the Moon, which we are treating as constant – though it is not).

178

Purpose : To use a telescope to observe and time the transit of the Moon and use basic algebra and geometry to determine the diameter of the Moon.

Materials :

- Telescope,
- Eyepiece,
- Moon,
- Timer,
- Slide Rule

Procedure :

1) On a clear nearly full Moon night (can be during the day as well) take out and set up your telescope to observe the Moon.
2) As with all Activities involving viewing objects, do not misuse the telescope or endanger your eyesight in any way.
3) In this case, use a low power eyepiece, such as 18 mm – 20 mm – or 25 mm.
4) Focus the Moon into view and move the field of view so that the Moon has either its preceding 'right' edge at the edge of the field of view OR the trailing 'left' edge just on the edge of the field of view.
5) In either case, you are to time the total amount of time that it takes the Moon to either leave or enter the field of view.
6) Note : You are not measuring its transit time across the field of view. It is only for it to enter or leave the field of view. In essence, we want the time it takes only the Moon to move past a point.
7) Do this for 3 trials.
8) Calculations :
9) First determine the Average Time for 'transit' into or out of view. (T_{ave})
10) Use this Average Time to determine the Angular size of the Moon. ($N°$)
11) (Note, we could measure this or use someone else's estimate, but the goal of the Activity is to use a time measure to determine this value)
12) From the Angular Size determine the Number of Radians for the Moon in size. (R)
13) Use the Radian value in conjunction with the known distance to the Moon to find an approximate Diameter of the Moon (Dia_{Moon})
14) Now determine the percent error for this value.

Data :

Trial	Time [T] (s)
1	
2	
3	
Average	

Calculations :

Be sure to use your Slide Rule!

Average Transit Time :

$$T_{ave} = \frac{\Sigma T}{n}$$

Time Conversion from Sec into Hours :

$$h = \frac{T_{ave}}{3600}$$

Conversion of Time into Degrees :

$$N^o = 15^o * h$$

Conversion of Degrees into Radians :

$$R = \frac{\pi * N^o}{180}$$

Diameter of Moon Calculation :

$$Dia_{Moon} = d * R$$

Percent Error :

$$\%E = \frac{[\,Actual\text{-}Measured\,]}{Actual} * 100\%$$

<u>Value Measures that may be useful :</u>

Distance to the Moon (d) :

        2.39×10^5 mi
        3.84×10^8 m

Actual Diameter of the Moon :

        2.160×10^3 mi
        3.475×10^3 km

<u>Conclusion :</u>

How well do your calculated measures of the diameter of the Moon match the accepted value and why do you think they are different?

Activity #19
The Size of the Sun, Sunspots and the Slide Rule
Grade Level : High School
Math Level : Calculating

The use of proportions allows one to calculate the size of or distance to an object in question. This is because of the principle of similar triangles.

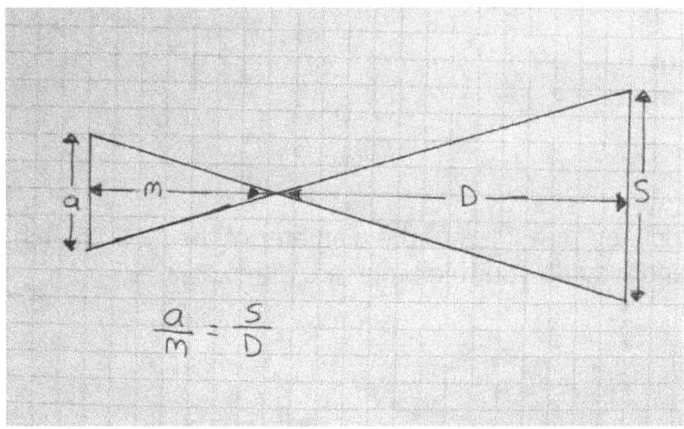

Interestingly this idea applies at all scales, from the microscopic to the solar. For example, the first determination of a distance to a star was done in a manner employing the knowledge of triangles, angles and known distances (such as Earth's orbit) to find the apparent movement of nearby stars as compared to far more distant ones.

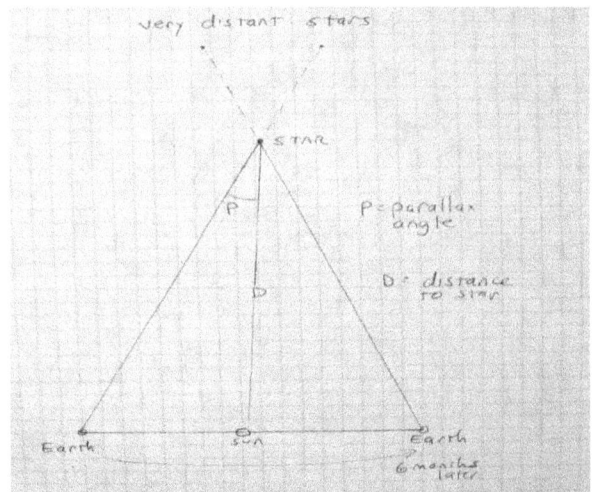

Our Activity will involve making a simple projection device for the Sun, measuring its apparent diameter and the distance from our pinhole projection source to the screen and knowing one fact, the actual distance to the Sun, we can then find the diameter of the Sun.

Notice that we are employing the proportion once again in this case (as can be seen in a number of activities and one involving them as well).

$$\frac{\text{Actual Sun Diameter}}{\text{Actual Sun Distance}} = \frac{\text{Apparent Sun diameter}}{\text{Sun Projection Distance}}$$

Of course, one could say, how did we find the distance to the Sun. That process happened in the 1800s and was due to watching transits of the planet Venus across the Sun, measuring and timing it well, plus employing geometry yet again to find the distance we are from our star, the Sun. From there, using Kepler's Laws it was easy to find the other distances of the planets both from the Sun and from each other.

Purpose 1 : To use projection method to create a measureable image of the Sun at a known distance and knowing the distance to the Sun and using proportions determine the size of the Sun.

Purpose 2 : To use a projection solar telescope and create a large enough image of the Sun where its apparent diameter can be measured as can the size of any sunspot surface features so that proportions can be used to find the actual size of the sunspot.

Materials :

- Meter Stick,
- thick Cardboard Card (2),
- Aluminum Foil,
- Tape,
- Pin,
- Pencil,
- Tape,
- Graph Paper,
- Ruler,
- Slide Rule

Note : Parental permission and supervision is advised. Do not look directly at or into the Sun. Do not use a scope inappropriately to damage yourself or it when using the tool to cast an image of the Sun. Safety also is in place for personal projection tools used and the same advice.

Procedure 1 : To use pinhole projection to measure the apparent diameter of the Sun and the projection distance in a proportion to solve for the diameter of the Sun.

1) NOTE : For any and all Sun Activities, **DO NOT LOOK DIRECTLY AT THE SUN** at any time.
2) Use a piece of cardboard cut into two pieces each 10cm by 10 cm.

3) In the center of one of them cut a square using an appropriate cutting blade that is 2 cm by 2 cm.
4) Tape a square of aluminum foil over the hole and tape across the face of it as well.
5) In the center of the foil use the pin and poke a small hole that just lets light through.
6) In the center of the other piece of cardboard tape down either graph paper of known spacing distances between the lines with additional marks at each millimeter or create a x-y graph with marks at every millimeter.
7) Attach a loop of cardboard to the base of the cardboard screens you have made so that the loop can go over and slide along a meter stick and the screens stand on the face of the meter stick.
8) Attach the pinhole screen card to one end of the meter stick and place the other at the other end and slide it towards the other one. Slide the screen card to about midpoint of the meter stick.
9) Now outside place the meter stick with the pinhole card over your shoulder and face away from the Sun. Let the meter stick lay on your shoulder as you move it with that arm's hand.
10) Watching your shadow line up the two screen cards.
11) Begin to move the projection screen card along the meter stick until an image of the Sun forms on it.
12) Mark with a pencil its diameter after centering it on the card. This will be the apparent diameter of the Sun (a).
13) Be sure not to move the screen card, since the distance from the pinhole screen card to the projection card is needed. (m)
14) With this data you can determine the size of the Sun (S). we will assume that the distance to the Sun is known ($D = 1.50 \times 10^{11}$ m).
15) Try this several times and be as accurate with your measurements as possible. Then compute the percent error with regards to the actual Sun diameter (1.4×10^{9} m).

Activity 2 : Determination of the Size of a Sunspot using a Solar Projection Telescope

Procedure 2 :

1) Use a Solar Projection Telescope for this activity (see photo)
2) Like any and all Sun Activities, DO NOT LOOK DIRECTLY AT THE SUN.
3) The Solar Scope allows for a projected image of the Sun onto a piece of paper.
4) Use a pencil and mark the sides of the disc of the Sun. This will be measured and used (sd) and is the apparent Sun diameter.
5) Mark and shade in pencil any sunspots visible. Be careful to mark the size noted by the projection since the goal is to determine its diameter. This is called (s).
6) Note in our calculations we will just use basic proportions and not employ spherical geometry corrections. Hence there will be some expected error, but realize that the closer the sunspot is to the middle of the disc, the better our estimation of its actual size.
7) From the first procedure, use the size of the Sun (1.4×10^9 m) as the Actual Sun Diameter (Su).
8) Calculate the estimated size of the Sunspot (Sp).

Data :

Activity 1 :

Trial	a (cm)	m (cm)

Activity 2 :

Trial	s (cm)	sd (cm)

Calculations :

Be sure to use a Slide Rule!
and note here that you will be using scientific notation.

$$\frac{a}{m} = \frac{s}{D}$$

$$\frac{s}{sd} = \frac{Sp}{Su}$$

Conclusion :

Be sure to do several trials and to keep safety in mind by never looking directly at the Sun. Then calculate percent error in these from the information given to check against.

Summary :

Extension Activity : Size of the Moon

Instead of measuring the size of the Sun in Activity 1, go out on a Full Moon night and do the same process with the projection system you have created. Here the distance to the Moon we will use is : 3.8×10^8 m and the size we are trying to obtain is : 3.5×10^6 m

Activity #20
Spectrum Analysis
Grade Level : High School
Math Level : Calculating

Many of us have read or heard about Sir Isaac Newton's work with the spectrum of light. He used a glass prism to separate light into its usual rainbow pattern we are all familiar with. What made his observation notable, though, was that he put other prisms in the path of each of the colors to find that you could not separate the light further. He rightfully concluded that white light from the Sun was indeed a combination of these colors, we often use the acronym false name of : ROY G BIV (Red, Orange, Yellow, Green, Blue, Indigo, Violet).

This spectrum of color is not the only components of the electromagnetic spectrum given off by the Sun, however. Many of us have heard about and encountered in one way or another other electromagnetic waves with names such as Ultraviolet and Infrared. To be 'ultraviolet' literally means 'above violet' while in the case of 'infrared' means to be 'below red'. From these descriptions it is easy to see that they are directly above and below the color spectrum that we can visually perceive.

This Activity will explore the Infrared and whether or not we can detect it through temperature measurements, though we cannot see it.

The Infrared portion of the spectrum today has many uses. We use infrared cameras in military, police, and security areas in order to locate people and animals in the dark due to their heat signature patterns. These cameras are used in heave smoke and fire situations to locate people and animals as well. Infrared imaging is used to examine heat loss from structures such as buildings and machinery. This is to study for defects, inefficiencies, and stresses. Infrared imaging data analysis is used in meteorology for looking at ocean currents and convection currents in the atmosphere. This is for weather both in the short term and long term. When used in imaging the Earth, it can be used to uncover ancient roadways and civilizations. While other analysis of the Earth data can reveal minerals as well as vegetation pattern. It is used in medicine to examine the body in a non-intrusive way. In Astronomy infrared imaging is used to analyze various objects and phenomena such as being able to penetrate clouds of dust to examine cradles of star formation.

The beginning of the journey in the Infrared can be traced back to Sir Frederick William Herschel (1738 – 1822), who was an astronomer who worked with his sister Caroline making catalogs of celestial phenomena such as nebulae and double stars and who also discovered the planet Uranus (1781). In 1800 he decided to follow up on Newton's spectrum work and was interested in how much heat energy did each of the colors of the spectrum contain. Much like our Activity, he used a glass prism to separate the spectrum into its respective colors and used thermometers with blackened bulb ends to see the effect of each portion of the spectrum on the change in temperature in a given time. To have a control in his experiment, he placed one outside the range of the visible spectrum. He found that each of the colors did indeed change the temperature more so than the control as would be expected. However there was an interesting pattern that emerged. From violet to red the temperature kept increasing and more surprising was when he placed the control thermometer below red it was the warmest of all and had the greatest change in temperature. In essence, he had uncovered the infrared portion of the spectrum. He first called these invisible rays 'calorific rays' from the Latin word for heat. Today, of course, we call them Infrared Rays. His further experiments should that these rays had the same characteristics of other waves – they could be absorbed and transmitted, they could be refracted, and they could be reflected.

Our goal is to do a similar experiment to Herschel's and examine the infrared portion of the spectrum as well. Enjoy :)

Purpose : To use basic materials to analyze the Sun's spectrum of light while examining the possibility of other electromagnetic waves passing through a prism from the Sun that are not visible (in this case infrared) and to measure the rate of change of temperature each portion of the analyzed spectrum causes.

Materials :

- Glass Prism (only glass, plastic is not effective, about 4" or more),
- Cardboard Box (best size is box size for photocopier paper),
- Alcohol-type Thermometers (up to 5 is best for the Activity),
- Black Paint or Permanent Black Marker,
- White Paper,
- Stopwatch or timer (measurements every 2 minutes),
- Graph Paper,
- Scissors and/or cutting tools for the cardboard box,
- Tape,
- Stack of thin books or magazines (for propping the box with regards to the Sun),
- Pencils, Rulers, other (for propping up thermometers in box to read),
- Table (card table for outdoors),
- Slide Rule

Note : This Activity is conducted outdoors and requires as clear of a sunny day as possible for best results.

Note : Do not look at the Sun. Be sure to have parental permission and supervision in the set up and in the Activity. For example, cutting the box for fitting the Prism will most likely require parental help and efforts.

Note : Though the number of Thermometers seems quite large and may be a problem in obtaining, one could still use as few as three if needed. See Set Up for more information.

Set Up :

1. The basic conclusion to the Set Up is as follows :
2. In using the completed spectrum device : It is best to have a table outside with the box on it.

3. The following gives general directions that you need to create a strategy for in accomplishing. It is best to refer to the photo below the Set Up for reference.
4. In the upper edge of the box on one of the short sides is a place cut out to fit the Prism in a snug fashion (can use some tape to help hold it in place).
5. It is best to have the prism towards the corner of this short side since you need to fit the thermometers inside the box and have the bulbs of the thermometers in the spectral light.
6. It is best to test the prism to see how it needs to be seated so as to allow light to pass through. Note that as time changes, you will have to change the angle of the prism since the Earth is rotating and the Sun's position is therefore changing.
7. Inside the box on the table or box bottom (depending on whether the box is inverted or not) are the thermometers all parallel to each other and each set in the spectrum portion under investigation. If it is not inverted, then it must have a lid since you want only a portion of sunlight coming from the prism to strike each of the thermometers.
8. On one of the long sides of the box is a hole cut in it to view the thermometers so that they can be read (may need a flashlight). Recognize that it is best to have the thermometers slightly angled so as to be readable (not too sharp of an angle). (See photos of drawings).
9. Note the size of holes depends on your arrangement and size of the prism. If the cardboard of the box is not too supportive of the prism, be sure to employ a small amount of tape (that is not across the place where the light is coming through). Other suggestions might be rubber or plastic stoppers or other pieces of cut cardboard or folded index cards cut in small enough pieces.
10. The thermometers have their own preparation. They need to have their bulbs blackened in appearance to facilitate absorption of light. This is done by either the black paint (can use black nail polish) or permanent black marker you have.
11. Note : The hole to see the thermometers is best to make it like a window that can open and close between readings (just cut 3 sides and fold). Recognize that you will always have to conduct the experiment at a similar time of day when first testing the box if your windows are too small in width.
12. In the completed set up, the bottom of the box should be lined with white paper.
13. An important decision comes from what part of the spectrum is to be analyzed, which depends on the quantity of equipment available – namely the number of thermometers. Clearly it is best to have one that will be in the Infrared region and in any good experiment it is best to have a Control to measure against for comparison which is the thermometer at the far end of the box out of the spectrum, but it is also good to have at least a third thermometer for at least one of the color portions of the spectrum (say red or blue, for example). – Realize though one can conduct the experiment a number of times in a short time frame and do different portions of the spectrum with your given supplies.

14. It is best to experiment and test the box and complete set up several trial runs without data to have a good idea as to a plan to conduct the experiment and record data. For example, there will be a considerable effort to find just the right position for the prism so as to catch the sunlight at its given angle (due to time of day) in order to project a spectrum on the floor of the box. – Also determine what is best to use to prop up the thermometers to read them easily, as well as slightly propping up the box when it is facing the sun in order to facilitate the sunlight passing through the prism. Essentially it is practice, patience, and focus. Enjoy :)

Photo of Set Up :

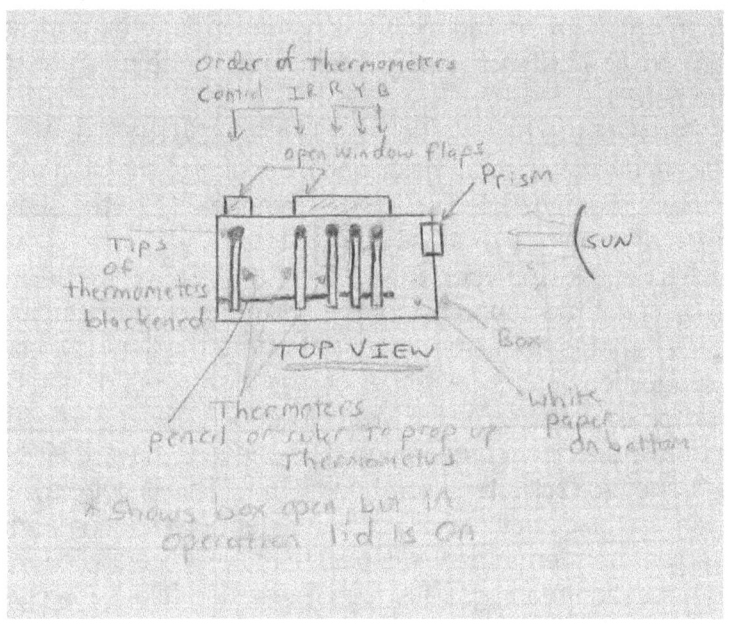

Procedure :

1. At this point you have read all of the materials, even the Procedure at least one time through. You have constructed your box to use a prism to separate the sun's light into its spectrum and have tested it from the Set Up section.
2. Always be sure to have parental support, permission, supervision, and help whenever necessary. Do not look into the Sun. Be sure to be careful when in the Sun, such as wearing sunblock.
3. To initiate the Activity, have all the materials set up and ready to go – sunny day, table, spectrum analyzer box all set up.
4. At this point though you have put the thermometers in the box, it is entirely sealed (that is to say, block any incoming light) to allow the thermometers to reach an equilibrium temperature with the outdoors. This may take 10 or more minutes.

5. Record the initial temperatures of each of the thermometers (they may be slightly different) on separate scratch paper, calculate an average and place this answer in the Initial Temperature reading space on the Data Table.
6. Set the box so that the incoming light produces the spectrum you want and need to place your thermometers where they belong. Record each of the thermometer's temperatures separately for time 0 in the data table.
7. Start your timer and every two minutes record the temperature reading from each of the thermometers in the data table. Do this for at least 6 minutes, 10 is preferable.
8. With all of your readings accomplished, you can put away your spectrum analyzer device.
9. Calculations :
10. For each of the Thermometers separately determine the Average Temperature for a given portion of the spectrum, or the Control.
11. Graph Time (x-axis) vs. Temperature (y-axis) points for each of the thermometers separately.
12. For each thermometer data set of points, draw a best fit line through those points (recognize it will intercept the y-axis at a point and not 0). There are as many lines as there are thermometers.
13. For each line, determine the slope of that line.
14. For comparison, look at the various colors studied along with the infrared (IR) and the Control slopes and note any conclusions you might reach. (Which is rising at the fastest rate?).
15. As with most experiments, it is oftentimes best to repeat this study for comparison to other days.

Data :

Initial Temperature of Thermometers (Average) in shade for 10 min. _____

Part of Spectrum		Temperature at	This time :		
	0 minutes	2 minutes	4 minutes	6 minutes	Average Temp.
Blue					
Yellow					
Red					
Infrared					
Out of Spectrum - Control					

Calculations :

Be sure to use Your Slide Rule! :)

<u>Slope :</u>

$$m = \frac{\Delta Y}{\Delta X}$$

<u>Average :</u>

$$Xave = \frac{\Sigma x}{n}$$

Conclusion :

What do your results show you about different parts of the spectrum that we can see in terms of the rate of change of temperature per unit time? What would it mean if a given part of the spectrum warms something up faster?

What did your experiment show for areas outside the visible spectrum? Can you conclude that there are regions beyond the visible – if so, how?

Activity #21
Calculating the Solar Constant
Grade Level : High School
Math Level : Challenging

The Solar Constant, Solar Energy and the Slide Rule Activity –

The Sun is the home star of the Earth and the whole of the Solar System. It is 1.4×10^6 km in diameter (approximately 109 x Earth's diameter), hence its volume could hold over 1 million Earths! The mass of the Sun is approximately 2×10^{30} kg (some 330,000 x that of the Earth).

Despite this the Sun is fairly average amongst the family of stars in the cosmos being classified as a G2V type. (Class ranges : O,B,A,F,G,K,M with each having 10 subcategories and the V is a Main Sequence star meaning it is in the prime of its life and generates energy from fusion of hydrogen into helium).

The Sun is composed primarily of Hydrogen (73.5 %) and Helium (25%) with the remainder being many common elements such as Oxygen, Carbon, Iron, Sulfur, and Nitrogen.

The Sun is not only radiant in visible light spanning the spectrum (ROYGBIV) as uncovered and described by Newton as being the components of white light) but also emits radiations at many other wavelengths, such as Radio Waves, X-Rays, Infrared Radiation, and Ultraviolet Rays.

In this Activity we examine the Sun's energy output directly.
Where does the Sun get all of this energy? Like other main sequence stars, the Sun has nuclear fusion taking place in its core. In this process in the 17 million degree core the Sun converts 630 billion kg of Hydrogen into 625.7 billion kg of Helium (a difference of some 4.3 billion kg of mass loss) per second. This results in a power of 3.9×10^{26} W given off by the Sun! In time through the processes of conduction, convection, and radiation, the energy reaches the surface in some 20,000 years and is emitted into space as electromagnetic radiation. The photons (all of which that move at the speed of light 3×10^8 m/s) that reach the Earth (some 1.5×10^8 km) take over 8 minutes of time.

The importance of the Sun's energy cannot be overstated. It is the primary reason that the Earth is at the temperature it is at. The Sun drives our atmosphere and weather systems. Hence it is the energy source for all wind generators as well as ultimately those that use water to drive turbines since it is the primary energy source of the water cycle on Earth. All plants use it through chemical reactions involving chlorophyll to convert water and carbon dioxide into sugar molecules which enables plants to grow and act as the pivotal base of the food chain for all organisms on Earth. This means that all Sciences (Physics, Chemistry, Biology, Astronomy, et al) have direct and minimally indirect relation to the Sun and its effects on the Earth. Today there is a new look at the Sun and its power and its importance to our needs.

Like all measures in Science, the energy given off by the Sun is just one of those quantities. From measurements a term has been developed called the Solar Constant. It is a measure of the flux (the amount of incoming solar electromagnetic radiation per unit area that is incident on a plane perpendicular to the radiation at a distance of 1 AU (astronomical unit)). This measure is all forms of energy coming from the Sun and is not just visible light.

The current measured value of the Solar Constant is 1.366 kW/m^2 on the average (note it does vary with our place in the solar system since at times of the year we are closer while other times farther away from the Sun).

This value does and has changed over time. The reasons for this are not fully known. The science of the sun is multidimensional (studying mathematical models of the interior of the Sun, helioseismology, solar sunspot activity, the mystery of the missing neutrinos, and explanations of the superheated corona, and many other areas) and far from complete. Even in 2010 it was found that the Sun's surface can have cascading fluctuations in the magnetic field causing massive solar storms – none of which were known of or had been predicted before. This also includes coronal mass ejections which even caused an electrical power outage in Canada in 1989 since it affected the Earth's magnetic field which in turn affected transformers and power lines.

When it comes to our measure of the Solar Constant, we are only looking at visible light and none of the other electromagnetic forms of radiation. Also we have to do this from the surface of the Earth instead of in space. This means in the Activity we are leaving out many of the other energy forms, but take note that the majority of the energy is in the visible light form, so our measured approximation is a good one. Our Activity also has to consider the fact that not all of the light reaches the Earth's surface (where our measurements are made) as well as the transparency of the atmosphere (is it hazy or not due to water molecules and/or dust) and where the measurements are taking place (the angle of latitude where one is at will affect the amount of sunlight that reaches the surface since it must pass through different amounts of atmosphere during the measurements). All of this is simplified by a 'constant' factored into the Activity to compensate for these things. More precision can be found by internet research for formulae factors to incorporate as well as more precise equipment and measurements.

Purpose : To determine the daily amount of solar energy that reaches the Earth's surface through the properties of the specific heat of water, the water's temperature change, measurement, and calculation.

Purpose : To determine an estimated value of the Solar Constant.

Materials :

- 2 Large Styrofoam cups,
- Water,
- Measuring Cup,
- Black Paint,
- Thermometer (lab quality),
- Clear Plastic Wrap,
- Flat Surface (large plate or pan or TV tray),
- Ruler,
- Paper,
- Tape,
- Clock or Stopwatch,
- Alternative Method Tools : Multimeter & Solar Cell,
- Slide Rule

Procedure :

Set Up & Needs :

1) The Radiometer (our solar energy collection device) is the two nested Styrofoam cups with the inside one having its entire interior painted black (be sure it is dry for the experiment).
2) The main need is a sunny day and the experiment is best ran in the middle of the day (between about 11 AM up to 2 PM).
3) You determine the angle to place your Radiometer at by placing the nested cups on a flat surface and elevating it so that at the time of the activity with the Sun shining on it there will be no shadow cast by the cup on the water in the cup. That is, it is directly facing the Sun.
4) For those who like numbers in this case, the angle at which the board will be to the level ground will be the same as the latitude of the location of the activity.
5) In the process of this set up determine how best to create an angled surface. A suggestion is a wood board or plastic TV tray (try not to use metal surfaces) as supported by books. Be sure to test this outside. You want the cup to hold in place, tape helps this.
6) On your angled surface place one of the Styrofoam cups and see how much water can be put in without spilling. The goal is to fill it in a way and at that angle so there is no shadow of the cup on the water's surface and the water does not spill out. (about 2/3 is good typically)

7) Measure this useful amount of water with a measuring cup. This will be the amount used in the Activity (V).
8) Note that the Thermometer will be placed in here too, so take that into consideration.
9) Always employ safety in all aspects of your Activity – do not look directly into the Sun.

Activity Procedure :

1) About 30 minutes before the Activity fill the nested cups with a measured amount of water (V) as determined from the set up above.
2) Note it is best to use room temp water neither too cold nor hot).
3) Place them on the angled surface outside facing in the general direction of where the Sun will be in 1/2 hour from now. Do not take the temperature yet. The goal is to have the water be at the same temperature as the surroundings.
4) About 5 minutes to measuring time cover the cup with the plastic wrap. At this time insert the thermometer too but still no readings.
5) It is best to insert the thermometer and place it in a manner for easy reading (to be seen from the side so as not to block the Sun striking the water's surface).
6) When the Radiometer has been in place for 5 minutes take your first temperature reading (T).
7) Record temperature readings every 2 minutes for 30 minutes (or at least until there is a temperature change of about 5°C), ∆t is all of the seconds elapsed.
8) Note that you may have to slightly shift the platform with the Radiometer on it to keep it in the sunlight since the Earth is spinning and the Sun therefore 'moves' across the sky.
9) When done with the measurements, carefully remove the plastic and thermometer.
10) Use the bright-colored paint (white, yellow, or silver) to mark a line on the interior of the cup at the water line all the way around).
11) Now you can dump out the water.
12) Use scissors to cut the interior cup down to the line you have drawn.
13) Invert this angled cut surface onto some paper (it can be regular or graph paper) and trace carefully the edge of the cup as close to the inner lip as possible without distorting the cup.
14) With the traced ellipse on the paper use a ruler to measure in centimeters both the long and short axes of the ellipse (A, B).
15) Calculations to Perform :
16) Though the Slide Rule is a recommended tool, all of these calculations can be done with a regular or scientific calculator. Some scientific ones even have built-in averaging formulae. For those who like spreadsheets, the data can be typed in and the formulae then also be typed in its own cell where the formula references each of the measured variables in their respective cells.
17) Use the Ellipse formula to determine the Area of the Water's Surface. Convert this answer to m^2.
18) Calculate the amount of mass in grams, then convert to kilograms, of water used. Note the volume should have been read in mL and 1mL of water equals 1 g of mass of water.
19)

20) After determining ΔT, find the Experimental Solar Constant (Q) by plugging in all of your measured variables.
21) With the determined Solar Constant, find the percent error from the Known value of it. Compare this value to the corrected value in the table of information below.
22) Using the measured Solar Constant from the Activity, determine the amount of Power [N] (J/s or watts) reaching the Earth's surface itself.
23) Now determine the total Power [P] of the Sun's energy that reaches the Earth's orbit.
24) Since we have determined the total power (P) that reaches the sphere of Earth's radius we have indirectly found the total Power radiated by the Sun. How does this value compare to the known expected value for the incoming solar power ?
25) Alternative Comparison Method :
26) Use a Solar Cell (Photovoltaic Cell) connected to a Multimeter :
27) When angled at the Sun (be sure to not be behind glass indoors), measure the Voltage (V) with the Solar Cell (most are in the neighborhood of 1.5 V),
28) Now reconnect the multimeter to measure milliamps (mA). Depending on where you live, the size of the solar cell, the overall transparency of the atmosphere this value can possibly range to over 250 mA (be sure to note the range capability of your multimeter and have it set correctly for a reading).
29) From the Voltage and Amperage readings, calculate Power : $P = V*I$
30) Measure the surface area of the solar cell actually collecting sunlight (look carefully at the solar cell to find the area actually receiving and processing the light). Convert this measurement into square-meters.
31) Like the prior calculation for the Solar Constant divide the Power determined from the electrical readings by the area of the solar cell and as before factor in your correction factor (k = 0.5 in the denominator, which in turn is really multiplying it by 2 in the numerator) for you Experimental Solar Constant Value (Q).
32) This value is good for comparative purposes and an alternative method to the aforementioned activity!

Data :

Volume of Water : _____ mL

Mass of Water : _____ kg

Ellipse : A : _____ cm
B : _____ cm

Time (in Minutes)	Temperature (°C)
0	
2	
30 (?)	

197

Calculations :

Be sure to use your Slide Rule!

Constants to be used :

$1 \text{ kg} = 10^3 \text{ g}$

$1 \text{ km} = 10^3 \text{ m}$

water specific heat capacity (c) of : $1 \frac{Cal}{g*°C} = 4.186 \frac{J}{g*°C}$

for calculations use $4.2 \frac{J}{g*°C}$

Whole Surface Area of Earth : $5.1 \times 10^8 \text{ km}^2$

Half of Earth Surface Area (SA) : $2.6 \times 10^8 \text{ km}^2$

Earth-Sun average distance (R): $1.5 \times 10^8 \text{ km}$

Correction Factor for Solar Constant Formula   (K) : 0.5
(note this can range from 0.05 to 0.95)

Formulae :

Mass of Water : (1 mL = 1 cc, 1g H_2O is 1 mL)

$$M = D*V$$

Area of Ellipse :

$$W = \frac{\pi*A*B}{4}$$

A & B are the measures of the axes

Solar Constant

$$Q = \frac{m*c*\Delta T}{K*W*\Delta t}$$

Power Received by the Earth on its Sun-facing Surface :

$$N = Q*(SA)$$

Total Power radiated by the Sun :

$$P = 4*\pi*Q*R^2$$

Percent Error : (use the values noted here)

$$\%E = \frac{[\text{Experimental Value-Accepted Value}]}{\text{Accepted Value}}*100\%$$

Constants to compare to :

Actual Solar Constant : 1.366 kW/m^2
Corrected Amount received at Earth for Solar Constant :
        342 W/m^2
Total amount of Solar Power reaching Earth's orbit :
        1.74 x 10^{17} W
Energy per unit time & area received at Earth :
        1.96 cal/(min*cm^2)

Conclusion :

How well did your values match the expectations of the Solar Constant and if not, what things affected your measures?
In trying the Alternative Method, how do the values compare?

Solar Energy Comparison and Determination with a Slide Rule Activity
Grade Level : High School
Math Level : Challenging

Consider for a moment – how does one measure light? It is not a distance, mass, or a temperature. It is, instead, its own quantity and referred to as Illuminance. It is the total luminous flux per unit area. Luminous flux is akin to the 'amount' of visible light present while illuminance is a measure of the intensity of the illumination on a surface. The SI units are measured in lux (lx) or lumens per square-meter. One lux equals one lumen per meter-squared. (An interesting fact is that lux is both the singular and plural of the word).

The full Moon on a clear night is about 0.27 lux (it can be up to 1.0 when overhead in tropical latitudes). A very dark overcast day can be 100 lux. Ordinary office lighting is 320-500 lux on average, while sunrise or sunset on a clear day is around 400 lux. Full daylight and not direct sun can be from 10,000 to 25,000 lux.

Another term that is similar, but not the same, is lumen. Lux takes into account the area over which the luminous flux is spread while lumens do not take into consideration the area. Another term used is footcandle, where 1 footcandle is approximately equal to 10.764 lux. This one is more common in the U.S. than anywhere else and is used in constructed-related engineering.

Notice there has been no discussion of the most common term used in everyone's homes – wattage. When a bulb goes out, we look for a given wattage bulb – not one based on its illuminance. Yet look at the side of a light bulb case sometime. For example, a common pack of incandescent bulbs (now becoming a rarity) might say 75 watts for energy used and light output of 1170 lumens. Though it is this bright, realize that this is not lux and that will depend on the illuminated area.

It might seem from the information given that one could find the number of watts per lumen readily. This is not so. Different bulbs use different amounts of power and also give off different amounts of light. Consider LEDs, flashlight bulbs, regular incandescent bulbs, halogen bulbs and fluorescent bulbs. Incandescent bulbs generate about 12 lm/W, while fluorescent bulbs can be in the area of 60 lm/W. In any case, however, if we use the same light source in our Activity, then we can generalize and treat the conversion as linear and complete. The only problem, then, is to find a way to measure the amount of light given off by a source, such as a light bulb or the Sun even.

The device is to measure the intensity of light is called a photometer. They can be very sophisticated and used to measure : Illuminance, Irradiance, Light Absorption, the Scattering of Light, the Reflection of Light, Flourescence, Phosphorescence, and/or Luminescence. We will construct one that come from the 1800s and compares the light given off by two different sources when the light is viewed from a side-view while illuminating a set of blocks of paraffin wax. The light will enter the wax and with a reflective surface (we will use aluminum foil) it will illuminate the block of wax. Having 2 of these blocks back-to-back and each being illuminated by a different light source, we can move the source or our device so as to effect the distance between the source and our photometer. When they appear equally bright to our eyes, then we can use the measured of wattage and distance to see if the system does indeed work. If one of the sources is the Sun while the other is a known bulb of known wattage, we can calculate the Luminosity of the Sun!

This is because a light source emits a given quantity of energy per second in its usual forms of electromagnetic radiation (visible light amongst them). Luminosity (L) is the amount of energy (measured in Joules) per second or Watts. The given brightness of a source is this divided by the area that is illuminated, which is spherical ($A = 4*\pi*r^2$). The ratio ($\frac{L}{A}$) for a given source can be set equal to that of a second source (like in our Activity) and we can cancel out the $4*\pi$ (since it is common to both sides of the equation). This shows that brightness depends on the inverse-square of the distance of the source (see Inverse-Square Law Activity). Since Luminosity (L) is the wattage (J/s) for a bulb we can compare it to any othe bulb or the Sun for that matter. Also the ratio ($\frac{L}{A}$) is the apparent brightness of the object and it has a direct connection to the actual brightness of the object. Using the Stephan-Boltzmann Law we can then find the temperature of the source via this relation.

We can also find the amount of energy given off by the Sun in the Solar Constant Activity which relies on the heating of water by the Sun as well for comparison.

Note : This is an Activity that requires the creation of the measuring device to do the work. Also it is important to have parental permission and supervision. Parents should be actively involved in the preparation, creation, and operation of the Activity. Always employ safety in using any and all materials. Do not look into the Sun or the lights, do not touch light bulbs, do not touch or handle hot objects or materials, keep hot lamps from contact with objects, and the like.

Purpose : To compare the luminosity of two bulbs (60 W & 100 W) through the proportions of luminosity and distance-squared

Purpose : To compare the luminosity output of a 200 Watt bulb to the Sun's power output to determine the Sun's Luminosity and from it, the Sun's surface Temperature.

Materials :
- Electric Lamps (2),
- 60 W & 100W (2) & 200W Bulbs,
- Meter Sticks (2) or Measuring Tapes,
- Paraffin Wax Blocks (as many as 4, index card size max),
- Aluminum Foil,
- Shoe Box or similar Cardboard Box,
- Ruler,
- Cardboard Piece (about the size of an index card),
- Plastic Crate, Chair, Stool, et al for set up for Sun
- Window facing the Sun,
- A Sunny Day,
- Slide Rule

Set Up of Photometer :

1) The Photometer consists of a layered structure (see photo) composed of 2 outer layers of paraffin wax, each with an aluminum foil (shiny face out) behind the wax and in the middle a piece of cardboard.
2) Start with a wax block standing on edge,
3) Next have the aluminum foil face so that the shiny face is in the direction of the wax,
4) Place the cardboard piece next and wrap the aluminum foil around the edge of the cardboard so that it faces out from the other side as well,
5) Up against this other side with aluminum foil facing out as well place the next paraffin wax block.
6) It is held together by rubber bands (they should have about ¼ inch width so as to distribute their surface force and not cut in too deeply)
7) Place this into a box that lets it stand on edge. Put it in the middle of the box.
8) If it is not big enough, the photometer system will be ultimately placed up against one of the walls (the one where you cut an opening to view it) while in the open gap area block this off with still more cardboard so as to create two distinct 'rooms' where light will enter from the box ends but do not affect each other.
9) Trace on the outer side of the box a place where a window will be.
10) Remove the block and cut the box so that there is a window – trim as needed so that each block can be seen equally.
11) In the ends of the box, trace a rectangle that is 2" x 2" (size will depend on the size of the box) and cut these out.
12) The holes in the ends are to let light in.
13) When fully constructed your Photometer, you look from the side at the paraffin wax blocks and note the comparative brightness of each of the blocks.
14) Moving one of the known sources of light being beamed through one of the openings, you adjust its distance until it matches the other light source illumination through the other opening (the Sun or another bulb or other light source).

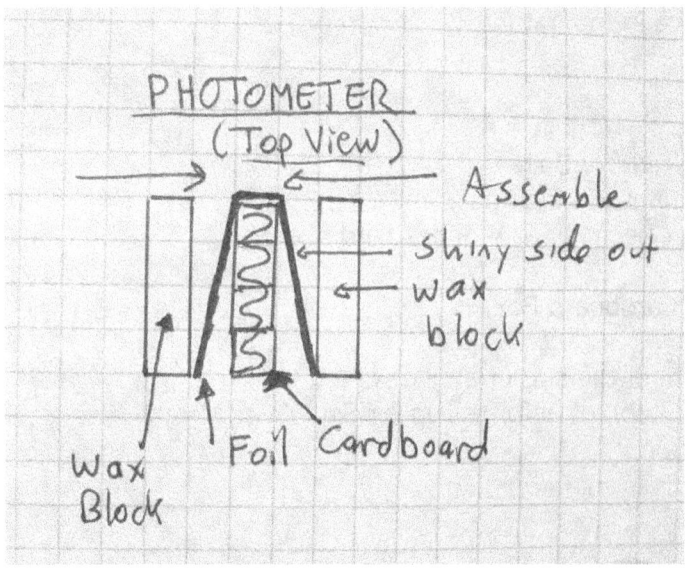

202

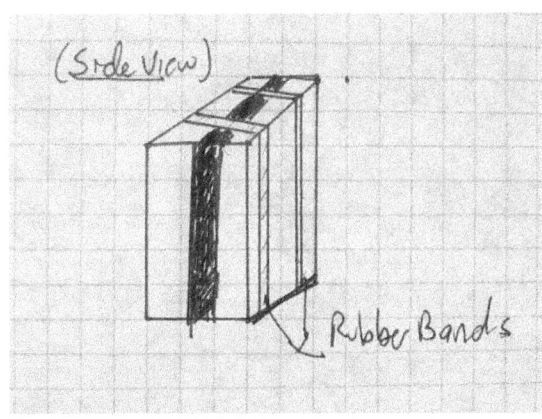

(Side View)

Rubber Bands

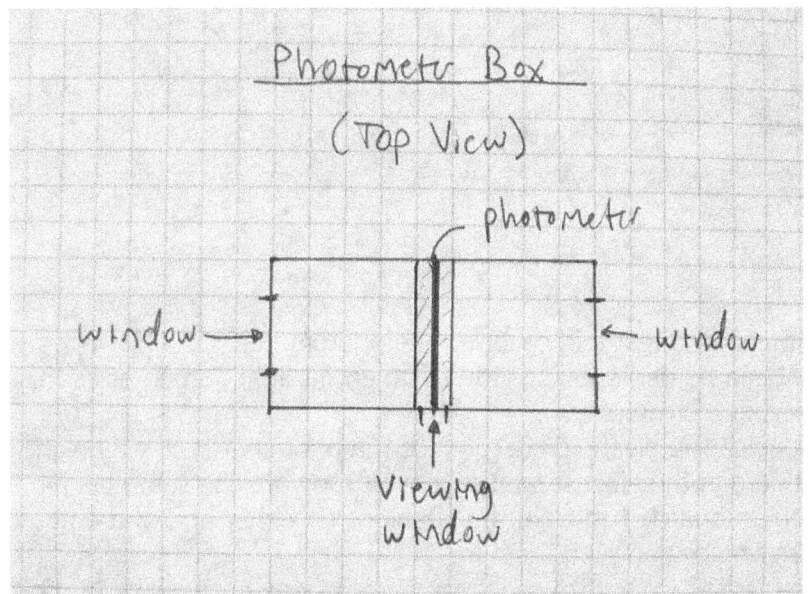

Photometer Box

(Top View)

photometer

window →

← window

Viewing
window

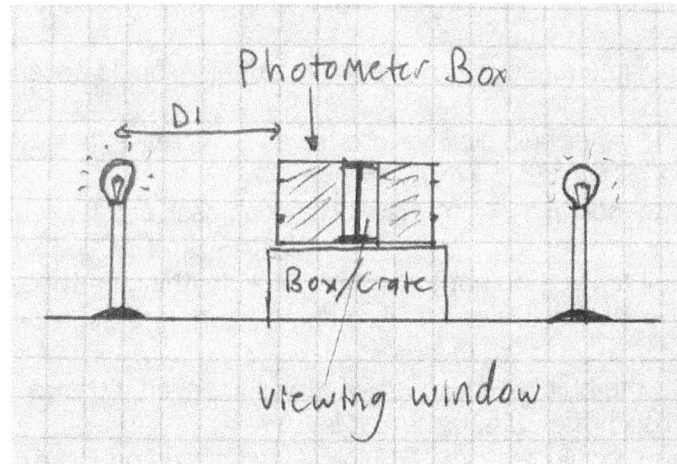

Photometer Box

D1

Box/crate

viewing window

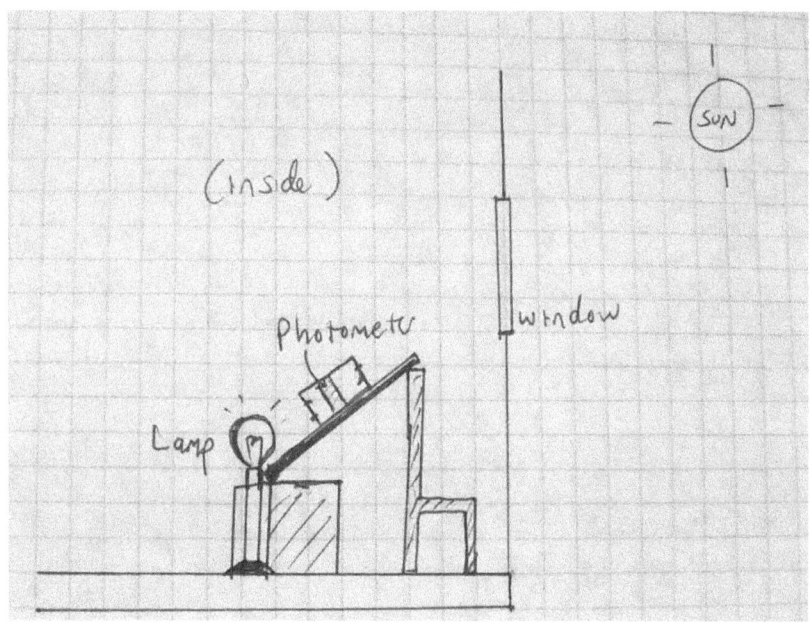

Procedure :

1) The first experiment is to test and calibrate the photometer.
2) Find a room that can be darkened so little to no errant light can enter as this affects the precision of your instrument.
3) Use a stack of books or a plastic crate to place the photometer at the height of the bulbs when they are in the electric lamps (best if they are the same height).
4) Use the lamps, placed each 1 m from the openings on either end of the photometer and having 100 W bulbs and plug them in.
5) Activate them separately to test to see if the paraffin wax on the side facing the lamp becomes illuminated when viewed from the side and none of the light affects the other wax piece.
6) Now activate both of them simultaneously. If ideally done, they should appear equally illuminated.
7) If one is brighter, move the other bulb slowly closer so as to increase its brightness. Adjust them until they are balanced.
8) Note it is a good practice to look away from the blocks, look back and let your eyes adjust.
9) NOTE : Employ safety at all times – do not touch nor place anything on the active bulbs and do not touch even when turned off. Also do not stare into the active bulbs nor the Sun at any time.
10) Note : When measuring to the bulb's (which is done for all needed distance measurements) Do Not Touch the Bulb, be sure it is Off!
11) To validate the system, you should retest with bulbs placed on the opposite sides and recheck the calibration.

12) If there are troubles, find ways to adjust the box, try different bulbs, check the distance, color the inside of the box black with flat paint or glued-on black construction paper, et al.
13) _Testing the Photometer._
14) As in the first exercise, on one side this time place a 100 W bulb in a lamp at a distance of 1m. (In other trials this can be any selected value).
15) On the other side, place now a 60 W bulb, first at a distance of 1 m.
16) With the lamps active, check the photometer and begin to move the 60 W bulb until the two wax blocks appear equally bright.
17) At this point, measure the new distance of the 60 W bulb from the opening of the box (do not touch the bulb and it needs to be off when taking measurements).
18) Record all the data in the Table below, the bulb's wattages (B1, B2), the bulb's distances (D1, D2). Note with distances add to it the distance between the box window and the aluminum foil on that side   (d1, d2). (Employ safety measuring to the bulb – be sure it is off and do not touch it).
19) Determine the Total Distances for Each (DT1, DT2) – this is the sum of the respective distance (D1 + d1, and D2 + d2).
20) Use the Luminosity-Distance Relationship with your Slide Rule to see if each ratio is equal.
21) If not, solve for the predicted distance that the 60 W bulb should have been (Dx). Use this as the Accepted Value and determine the percent error of your system (%E).
22) Time permitting, do this up to 4 times and be sure to reverse the box to see if this affects the outcome.

23)   The Luminosity of the Sun Test :
24) This next part requires a window to the outside on a sunny day –
25) The set up occurs near the window facing the Sun that shines in through a window.
26) On an inverted plastic crate place a board so that it acts like a ramp where the other end is on a sawhorse, chair or table.
27) You need to adjust this so that the photometer can set up on it – point it directly at the Sun. Seek stability for the system. Also find the right tools that allow for this. It clearly depends on the window's height, the other item's sizes, and the like.
28) Note that the photometer will be moved in this exercise, so practice this so all other items remain in place and do not become a problem.
29) At the opposite end set up the lamp with a 200 W bulb and plug it in via extension cord to an available outlet.
30) The bulb should be in direct line to the photometer window that will act as its receptor.
31) Note that the active bulb in this exercise will never be moved and only the photometer will move as needed – toward or away from it.
32) Viewing the illuminated wax blocks from the bulb and the Sun, move the box until they appear equally illuminated.
33) With that distance, turn off the lamp and pull the shade. Measure the distance from the box to the bulb (without touching it and be sure it is off) (D1) and add this to the distance from the window to the foil   (d1) – this is the total distance (DT).
34) Do this again starting with the box at a different place and take a new measurement.

35) Do this twice more with the box turned around and take the same measurements.
36) You now have 4 measures. Average them into an average total distance value (DT_{ave}).
37) Calculations :
38) Use the Luminosity-Distance Relationship to find the Luminosity (the Wattage) of the Sun! (Hint : You may need the Distance to the Sun for this). Note also the need to change units!
39) With this Luminosity, use the Temperature-Luminosity Relationship to find the surface Temperature of the Sun! (Hint : You may need the Sun's Radius for this).
40) From the known values, compute percent Error (% E).
41) If interested, convert the surface temperature of the Sun into Celsius and/or Fahrenheit.

Data :

Photometer Calibration with Two Different Bulbs :

Bulb 1 : _____ (W)
Bulb 2 : _____ (W)

Distance from window to foil (d1) : _____ (cm)
Distance from window to foil (d2) : _____ (cm)

Trial	Bulb 1 Distance [D1] (cm)	Bulb 2 Distance [D2] (cm)

Sun Luminosity Determination Measure :

Bulb Wattage : _____ (W)

Distance between foil and photometer window :

d1 = _____ (cm)

Trial	Bulb Distance [D1](cm)
Average Value (DT_{ave})	

Calculations :

Be sure to use your Slide Rule!

Luminosity-Distance Relationship Formula :

$$\frac{L_1}{4*\pi*d_1^2} = \frac{L_2}{4*\pi*d_2^2}$$

Note : Notice that $4*\pi$ will readily cancel out
Note : This can be used for 2 Bulbs or the Sun and a Bulb

Stephan-Boltzmann Law :

$$B = \sigma*T^4$$

Note : T is in Kelvin
Note : $s = 5.67 \times 10^{-8} \frac{J}{m^2*s*K^4}$

Relation of Luminosity and Temperature :

$$\frac{L_1}{4*\pi*R_1^2} = \sigma*T^4$$

Percent Error :

$$\%E = \frac{[\text{ Accepted Value-Measured Value }]}{\text{Accepted Value}} * 100\%$$

Converting Temperatures :

$$^\circ C = T_K - 273^\circ$$

$$^\circ F = 1.8*^\circ C + 32^\circ$$

Needed Values :

Note : The Wattage of the Bulb can be used for L, since the units of L are J/s or Watts

Distance to Sun from Earth : 1.5×10^{11} m

Radius of Sun : 6.96×10^8 m

1 m = 100 cm

Comparison Values :

Luminosity of the Sun : 3.9×10^{26} W

Surface Temperature of the Sun : 5800 K

Conclusion :

How well did your apparatus work when examining two different bulbs at measured distances? What was the percent error in this case?

With regards to the Sun, how close did your values come to the actual Luminosity and Surface Temperature of the Sun?

What factors could be affecting your measurements?

The Sundial is any device that measures time by the position of the Sun. There is no one origin of the device and they have been around for many centuries. With time, there have been different designs and even formulae to correct for one's latitude on the Earth.

The key to most of the Sundials is that there is a shadow cast by a Gnomon, typically a rod or angled piece that stands perpendicularly to the surface of the clock itself. The clock face has marked hours and other marks between the lines. There are types that cast a line of light or a spot where sunlight comes through a slit or a hole instead.

The basic types are Equatorial Models, Horizontal Models, Vertical Models, Pocket Models, and others.

Despite the type of Sundial, they must all have alignment with Earth's axis of rotation so as to have correct time. Often the Sundial must point towards true celestial north   (and hence not the magnetic North). Also the gnomon or Style as it is sometimes called will have the same angle as the latitude of the person using it.

The clocks tell the Solar Time, hence may have different values than that of the clock, if your area has daylight savings in place, for example.

To match the actual clock time, several corrections are needed : 1) Note that the orbit of the Earth is not perfectly circular and its rotational axis is not perfectly perpendicular to its orbit. This correction factor can be as large as 15 minutes and this is called the Equation of Time. 2) Solar Time has to be corrected for the longitude of the sundial relative to the longitude of the official time zone. Here a rotation of the hour-lines by an angle equal to the differences in longitude is needed. 3) the adjustment for daylight savings too.

Note that the vast majority of those bought for decorative purposes do not operate as clocks and are not fine-tuned to do so.

Sundials along with other instruments date historically to nearly as old as civilization itself, since time telling is so critical a feature. Devices were made not just for the Sun but also for the stars and their positions as well (Astrolabes and the like). These could be used for time telling for time of day, time of year, and hence the time of seasons and their changing, time for planting and sowing. It is also used to help determine place on Earth as well (latitude and longitude). This is helpful for navigation, map making, distance determination). So these devices were massive in scale as well as hand-held too.

As time went by, the need to develop mechanisms other than the natural universe for time telling became all the more important. With the advent of the steam engine and locomotives, the necessity of regulated time schedules became paramount. Also with the Industrial Revolution, the need for finer time measures, the work clock, became institutionalized. So Sundials became more decorative, but time is ever as important, and still solar based.

Our Activity deals with the human scale and has two types to examine. Each is personally made and used.

Purpose : To construct and use a home-made sundial to measure time and compare its level of accuracy to an actual clock through 2 (or more) different alignment techniques (daytime and nighttime).

Note : This same purpose can apply to a second design (as noted here) and then compare and contrast the outcomes of the two sundials.

Materials :

1) **Sundial #1 :**
2) Piece of thick poster board measuring 54 cm by 20 cm (or more),
3) ¼ in. dowel rod, length 50 cm,
4) Ruler,
5) Hole Punch,
6) Scissors,
7) Black ink pen (thick style) or fine tip marker,
8) Protractor,
9) Tape,
10) Slide Rule,
11) **Sundial #2 :**
12) Piece of thick poster board 30 cm by 30 cm,
13) Second piece of thick poster board 30 cm by 30cm,
14) Protractor,
15) Ruler,
16) Scissors,
17) Thin dowel rod 1/8 in or less, 50 cm long,
18) Glue,
19) Slide Rule

Set Up of Sundial(s) :

1) **Construction of Sundial #1 The Equatorial Sundial:**
2) Place the thick piece of poster board on the table and measure from one end along the longest dimension a mark 2 cm from the edge, followed by a mark at 20cm, then 30 cm.
3) Do this twice so as to make a line across the poster board at each of these marks.
4) Score the 20 cm mark line carefully (so as to not cut through the poster board) with scissors and a ruler.
5) Measure and mark the center of the first line that is 2 cm from the edge.
6) On the center point place the protractor so that it sits in the 20 cm area.
7) Trace its outline.
8) Knowing that 1 hour of time equals 15°, ½ hour equals 7.5°, and ¼ hour (15 minutes) equals approximately 3.8°,
9) Use The information to mark each 15 minute interval through the entire range of the protractor and connect each of the line.

10) It is probably best to have the hour lines as the thickest and darkest so that they are distinguishable.

11) Note that the base line will be written as 6 PM on the left hand side and 6 AM on the right hand side while facing the protractor image and the half-circle is pointed towards you.

12) Note that the line perpendicular to the base line is Noon.

13) If done correctly, reading from left to right with the semicircle towards you, the times read 6 PM, 5 PM, …, Noon, 11 AM, …, back to 6 AM

14) Poke a pin through the poster board at the midpoint through to the other side.

15) Flipping the poster board over, draw the same line at 2 cm from the top and reproduce the image of the protractor time as you did on the reverse in the same fashion, only now 6 AM is on the left and 6 PM is on the right.

16) Note, if done correctly, the 6 PM times (much like the 6 AM times) are directly on opposite sides of the board, much like a mirror image.

17) The reason for the double image is this – it depends on what time of the year you are using the sundial. The summer months will cast a shadow off of the gnomon (the dowel rod that will be put in place next) on the top surface, while in winter months, the underside will have the shadow cast on it by the gnomon.

18) Now punch a hole where you pushed the pin through with a hole punch.

19) Fold the poster board at the scored mark of 20 cm

20) Look up your Latitude (to the nearest $1/10°$ as this is needed to place the Gnomon (this is the dowel rod that will be the shadow caster for our Sundial).

21) First calculate (using a slide rule) the length of dowel rod needed to go through the hole to make contact with the poster board. This is denoted 's'. The formula is s = ——— . Here 'Θ' is your Latitude.

22) The amount that sticks back through can be left alone. Once in place secure with tape. You can check the angle with a protractor by placing the system on a table, near the edge and measuring the angle that the dowel rod makes with the poster board.

23) With the aforementioned steps the Equatorial Sundial is complete and ready for use.

24)   Construction of Sundial #2 : The Horizontal Sundial :

25) Like the Equatorial Sundial Model, use thick poster board.

26) On the poster board, place a protractor. The circular part of the protractor faces away from you.

27) Along the base, draw the base line, where 6 AM is to the left and 6 PM is to the right. The $90°$ mark is 12 Noon.

28) Instead of marking each $15°$ as an hour as we did before, here, since the Sundial model is flat, we need to alter the angle markings to account for the latitude we live at and the hour where the mark will go.

29) The formula to use is : $\tan(D) = \tan(t) * \sin(Θ)$.

30) D is the angle made with respect to the Noon time

31) t is the time in question turned into an angle from

32) $\Delta t = [$ Noon – time in question $]$

33) Note that 6 AM and 6 PM do not and cannot be done ($\tan 90° =$ undefined).

34) and where t = $(\Delta t) * 15°$ now expressed as an hour, where 1 hour = $15°$.   (Note : Each fraction of an hour is a decimal part of an hour and made part of the answer. For example, 0 min is $0°$, 15 min = $0.25°$, 30 min = $0.50°$, and 45 min = $0.75°$).

35) EXAMPLE !

36) Angle t for example for a time of 9:30 is Δt = [12 – 9:30] = 2 hours 30 minutes = 2.5 hours. Then t = $(\Delta t)*15°$ = 37.5°

37) Note this is only the angle and still needs the first noted formula which also requires Θ, the latitude.

38) Θ is the angle of latitude for your location.

39) For continuation of this example, we will assume a latitude of 40°.

40) To solve the angle we then do this :

41) $Tan(D) = tan(37.5°)*sin(40°)$

42) $Tan(D) = (0.767)(0.643) = 0.494$

43) First move the cursor to the tangent of the angle (Angle found on T then answer on D.

44) Move the C Right Index over this answer on D and read from the S Scale the sine of the other angle on C.

45) Move the cursor to this place to find the answer to the product on the D Scale.

46) To find D, read from the D Scale back to the T Scale for the angle:

47) D = 26.3°

48) Though tedious, it is good slide rule practice and great use of various scales in the process.

49) Back to the Directions in general :

50) One thing to be gained is that we need only do the times from 6 AM to noon, since they have complimentary times, hence angles past noon.

51) From our example, the complimentary time to 9:30 is 2. 5 hours past noon or 2:30, so Δt is the time on the clock past noon! In this case, 2:30 would have the same 26.3° angle as well!

52) We have to do this for the following sequence of times :

53) 6:15, 6:30, 6:45, 7:00, 7:15, and so on each 15 minutes up to noon to complete the cycle.

54) As with the Equatorial Sundial measure and mark each of the lines using a protractor, ruler and good pen. For the smallest time increments a mark may be sufficient and lighter lines for the half-hour marks may be considered, while the darkest lines are for the hour marks (for both Sundials).

55) With a second piece of poster board you need to cut a triangular gnomon from it.

56) Use a protractor and mark an angle that is equal to your latitude. Use a ruler to make a line across the entire poster board. Cut across this. This triangular piece is the gnomon (it should be the same size as the half-circle).

57) The gnomon is glued to the sundial with the point at the place where all the marks intersect, which is the focus or midpoint of the protractor image initially drawn.

58) The maximum angle is at Noon.

59) To help keep it vertical, use a thin dowel rod that is shorter than the gnomon (leave about 2 cm at top and bottom) and glue it along the gnomon on each side of it.

60) The Horizontal Sundial is now ready to be used.

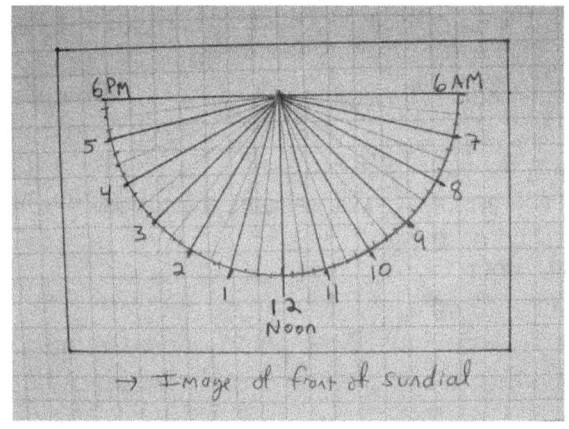

→ Image of front of sundial

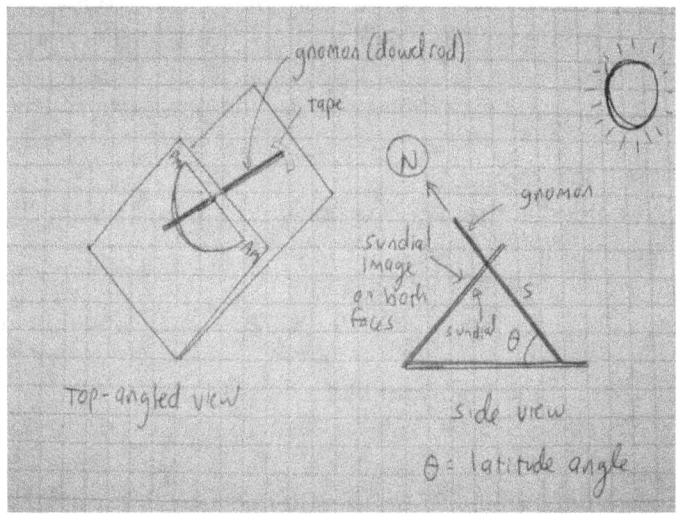

Top-angled view

side view

θ = latitude angle

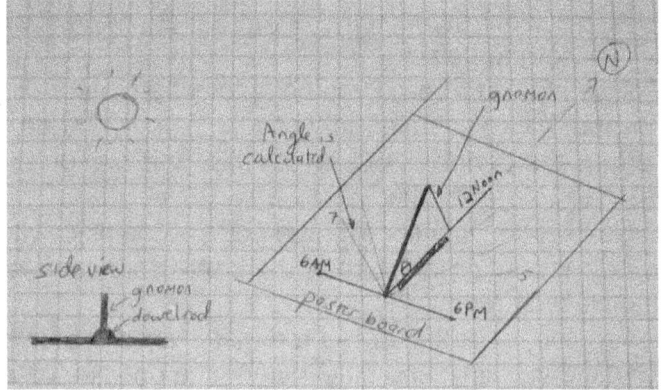

Procedure :

1) Before any measurements can be made,
2) You must Choose the alignment method for the Sundial and keep notes on the procedure you follow.

3) Alignment Method #1 Nighttime Star Alignment :

4) At night, have your Sundial out and use the North Star, Polaris, then end star in the little dipper asterism which is in the constellation Ursa Minor. And use this as 'true north' (it will be within 1°)

5) OR

6) Alignment Method #2 Compass Method :

7) Use a compass to find magnetic north and use this.

8) Note, with a little more precision and knowing the angle of deflection for your area (on outdoor camping maps) you can adjust this for differences.

9) You could always set the clock on a given day –

10) Once Noon on the regular clock, then adjust the Sundial so that there is no shadow and leave it in this position.

11)   Now With the Alignment Method in place :

12) Recognize that you have to use some sort of chalk marks or some method to set up the sundial daily (unless able to keep it covered and protected outdoors) so that it has the same positioning each day (and note, with each set up this will add to the error in the measurements).

13) Suggestion : A plastic box larger than the item can cover it, such as a Tote and should be enough protection.

14) Record the date, time of day (Actual Time) of measurement using the same clock, and the Sundial Time.

15) Note : That there are times that may have no values at all due to weather situations.

16) Note that it is best to choose minimally twice per day, if not as many as four times per day but it is best to be consistent with the times of day when you take your measurements.

17) Note : Do not always choose the hour for measures. Have at least one and up to three of the times be during a given hour.

18) An example might be : 9:00 AM, 11:15 AM, 1:30 PM, 4:45 PM

19) For any given day, be sure that in the case of your Sundial that your times measured are estimated to the nearest 2 minutes approximately (assuming that you have marked off 15 minute increments)

20) For each data point subtract the Actual Time from the Sundial Time.

21) Do these measurements for at least 2 weeks.

22) Note : For the Teacher : A good lab experience is to do measurements one week before the enactment of daylight savings time and proceed into a week after it has started or vice versa, so that the weeks straddle the employment or stoppage of daylight savings. Ask the students if their clocks are suddenly having troubles and why do they think that this is the case?

23)   Calculations :

24) Always be sure to use your slide rule!

25) Sum up all of the differences in time between the Sundial (T_S) and Actual Clocks.(ΔT)

26) Calculate the Average Number of Minutes Difference per Measurement (   )

27) Use this Average value to determine the difference of the data values from the mean and sum these values up.

28) From this determine the Standard Deviation of your Measurements.

29) Try another alignment method and record the data in the same manner as you did before.
30) Perform the same calculations and determine which of the two methods had the higher average and larger standard deviation. Why do you think this was the case?
31) Time permitting, use another Sundial or do it at the same time using the same alignment method and compare results of the two sundials noted here in this exercise.

Data :

Sundial Alignment Method : _____

Date	Sundial Time	Regular Clock Time

Calculations :

Be sure to use your Slide Rule!

Equatorial Sundial :

$$S = \frac{20 \text{ cm}}{\tan\theta}$$

Horizontal Sundial :

$\tan(D) = \tan(t)*\sin(\Theta)$

Can also use : $\log(\tan(D)) = \log(\tan(t)) + \log(\sin(\Theta))$

$\Delta t = [\text{Noon} - \text{time}]$

$t = (\Delta t)*15^\circ$

$Ct = \Delta t$ (Complimentary Time past Noon)

Data Calculations :

$\Delta T = \text{Actual Time} - \text{Sundial Time} (T_S)$

$$T_{ave} = \frac{\sum \Delta T}{n}$$

Statistical Examination of Data :

n = is the number of data values, ⁻ is the mean

Sample Variance

$$s^2 = \frac{\Sigma(T_S - Tave)^2}{n-1}$$

Sample Standard Deviation

$$s = \left(\frac{\Sigma(T_S - Tave)^2}{n-1}\right)^{1/2}$$

Conclusion :

How well did your clock work? Did one method work better than another for alignment – if so, why do you think so? What could you do to make the process better?

Activity #24
Graphically determining Kepler's 3rd Law
Grade Level : High School
Math Level : Challenging

Kepler's Law Calculations Activity:

This Activity begins with an essay I created discussing the importance of logarithms and their use by Kepler in uncovering natural science laws and rules we still employ today. With his 3rd law there is a way to uncover what he found to be true of the period of an orbiting body and its distance from the main body that it orbits using a slide rule and employing the L scale.

Kepler's contribution to Logarithms and the first application of them in Astronomy

Many a student and scholar has heard the name Johannes Kepler who is famously associated with laws bearing his name, Kepler's 3 Laws of Planetary Motion, but there is much more to this gifted mathematician, scientist, and astronomer than is summarized in a few pages of a descriptive astronomy course text. One of the key note features of his life is the contribution to logarithms as a math form to employ and its use in deriving his still used laws. The following explores Kepler and connects him to science and math as well as explores a way to examine his laws and any others for that matter in the classroom through logarithms.

Johannes Kepler (1571 – 1630) is called by many today as the last astrologer and the first astronomer. Along with that he has a number of remarkable firsts one might not know : he is first to investigate and describe the formation of pictures with a pinhole projection system, pave the way for image formation description with ideas of real, virtual, upright and inverted images from lenses and dealing with magnification, recognize and explain that depth perception comes from the use of both eyes, describe the process of our vision via refraction, formulate designs for eyeglasses for both nearsightedness and farsightedness, and explain how a telescope works. It was his works in Astronomia nova (the New Astronomy), Harmonics Mundi (The Harmony of the World), and Epitome Astronomiae which contains his laws that Newton read and acted as the spark to explore the more refined laws of motion to explain both celestial and terrestrial mechanics and leading to the description of the universal law of gravity which can be used to derive all of Kepler's Laws. In addition to these things and others is his derivation of logarithms purely from mathematics and their use in analyzing Tycho's data that was used to derive these laws.

A summary of Kepler's Laws :
1) The orbital paths of the Planets are Elliptical with the Sun at one focus.
2) A radius vector from the Sun to a given Planet will sweep out an equal area in an equal amount of time . (AKA the Equal Area Law)
3) The square of the Planet's Orbital Period is proportional to the cube of its semimajor axis distance. (This distance can be considered as the mean distance).

The importance of these laws, when taken in conjunction with Galileo's observations through a telescope in 1609, finally topples the long-held geocentric model of the universe in favor of the heliocentric model. Kepler's 1st Law eliminates the notion of circular orbits and smaller circles on these circles to compensate for errors in measurement. Kepler's 2nd Law for example removes the idea that planets move at uniform speed. For the Equal Areas in Equal Times to work for an Ellipse, the planet moves fastest nearest the Sun and slowest when farthest from the Sun. Newton's Law of Gravity gives the cause to this effect.

Kepler's first two laws were published together in Astronomia nova (A New Astronomy) in 1609 and had taken him years to arrive at by examining many mathematical models and restarting the process over and over. His dedication to calculation matching reality cannot be overlooked, though. His math models at one point only differed by 8 arcminutes (the full Moon is approx. 30 arcminutes) from Tycho's observations. He would not accept this. This process of focus on actual measurements and objective diligence is what we call the 'scientific method' today. These years of calculation was called by him his 'war with Mars' since Mars was his primary set of data under consideration. Had logarithms been available, these years of processing could have been done in hours by an individual.

Still motivated, Kepler felt that there must be a connection between all of these numbers. This third law was even more tiresome and was not published until 1619 in Harmonices Mundi. An interesting question arises, how did the notion of the square of one variable value for planetary data being equal to the cube of yet another variable seems surprising to say the least. To help frame this time, realize that John Napier of Scotland had published in 1614 his work on logarithms called Mirifici Logarighmorum Canonis Descripto. Napier is noted to have written in the instroductory comments to his own tables " a logarithmic table is a small table by the use of which we can obtain a knowledge of all geometrical dimensions and motions in space (from http://www.mathpages.com/rr/s8-01/8-01.htm p. 3) It is reported that Kepler had seen this work by 1616.(http://www.mathpages.com/rr/s8-01/8-01.htm on p. 2) Kepler's own interest in logarithms is even revealed in his 1621 publication on them. Further examination this history reveals the following, though: In 1620 in Kepler's Ephemerides he not only republishes his 3rd Law but also dedicates it to Napier (who had died in 1617). Further Kepler was working on the Rudolphine Tables and published with them in 1628 were not only the astronomical tables to predict star and planet positions were tables of 8 figure log tables as well as a response to the mathematican Maslin who admonishes Kepler for his use of this new idea in math since no one was clear on its validity. Kepler uses Euclid's Elements Book 5 to demonstrate clearly the mathematical validity of logarithms. Further the tables generated in the book upheld Kepler's ideas since they accurately predicted observational outcomes for many decades hence adding to the strength of the heliocentric theory. (A quick aside is the fact that in the time frame of 1620s to 1630 with 1620 and Gunter's Scale, and around 1622 with William Oughtred's creation of his Circles of Proportion otherwise known as the Slide Rule based on logarithms and their math laws).

It is here then that much credit should be bestowed upon Kepler for his dedication not only to science but to mathematics and the opening of the door to the world of logarithms which can be considered the prime power in calculation that comes to all parts of the now-emerging industrializing, voyaging, and investigating world. Logarithms as noted by Florian Cajori states : "The miraculous powers of modern computation are largely due to the invention of logarithms" (p. 1) which begins his work on the history of the prime tool to come from them, the Slide Rule in the book: A History of the Logarithmic Slide Rule and Allied Instruments (first published in 1910). All fields of science, mathematics, engineering have benefitted from them.

As an educator I am interested in sharing all of these ideas, but too much information, too many lectures can lead to a lack of interest and even understanding. Also it appears that Kepler's 3rd Law seems to come from nowhere and has no rationale whatsoever. Instead why not provide (or have access to) a table of basic planetary data such as their distances from the Sun and periods of revolution about the Sun. That is the basis of the following Demonstration of Kepler's 3rd Law Activity. This same idea appears in my book of activities to be done by using a slide rule. Accompanying the article is both the incomplete data table and one that is complete with the graph to illustrate.

Students can then be lead with as many directions as deemed necessary to investigate these numbers. First choose a planet to act as one to compare to. The Earth is a natural choice. Then put all of the data in terms of the Earth. The use of either log tables or the slide rule for this activity can be very useful. In the case of logarithms, since the value is a ratio, one only has to subtract the table values of numerator and denominator to find a solution to look up and put down the antilog of in the table. The level of the student will determine where to place the work here.

Next, take this relative table of values and look up their log values (or again use a slide rule). Since the Earth was chosen as the base to measure from, the planets Mercury and Venus will have negative values. The Earth will become the values of zero and zero since it is the log (1) for each of the values and will be the point (0,0) on the upcoming graph of this data.

Now graph the Log of the Orbital Period (Earth Years) versus the Log of the Semimajor Axis (in Astronomical Units). This can be done by hand or done in Excel for teachers wanting to employ computer technology. In fact the calculations can be done with Excel as well. I suggest the old-fashioned way. Some data points are Venus (-0.41, -0.211), Earth (0,0), Mars (0.183, 0.274) for example. Drawing a best fit line and computing the slope of this linear line by choosing two points reveals a value of 1.5. (Be careful with Excel as to which column is the Y and which is the X axis and the computation of the slope from this- my results yield 1.497 so at 2 or even 3 significant figures it is 1.5).

Since this the ratio of log of period (P) to log of semimajor axis (A) this means that the numerical ratio is 1.5 to 1 means 3 to 2 in terms of whole numbers. Use cross products and then raise each side to the power of 10 to eliminate the log notation, we find the cube of the semimajor axis is equal to the square of the orbital period ($A^3 = P^2$) just as Kepler had found!

The great thing with astronomy and physics are that there are a number of logarithmic relations that can be explored in such a manner, such as the relation of distance to apparent magnitude and absolute magnitude ($m-M = 5*\log_{10}(D/10 \text{ parsecs})$), star mass and luminosity ($L \propto M^4$), luminosity and radius ($L \propto r^2$), luminosity and surface temperature ($L \propto T^4$), mass and star lifetime (Star Lifetime $\propto M^{-3}$). In fact any and all data tables can be graphed in the log fashion employed here to uncover the relations between variables and derive the needed equations. This can help to develop math skills and connect them to science and technology. In conclusion, Kepler clearly has a place at the table of not just scientists, but also mathematicians and was critical in the realm of logarithms to their calculation, mathematical demonstration of, and practical connection of them to natural phenomena.

- This next activity is spaced-based. In Astronomy there are some useful general relations for bodies (stars and/or planets) that orbit each other called Kepler's Laws.
- You may recall the reference to him in the above article and his contributions to logarithms.
- I go through how to use the slide rule and reading the scales to find the various values, but you can go right ahead and use the table afterwards and find the log value for each period and distance value, plot these as points, find the best fit line and determine the slope which in turns yields the power relation of these variables.
-
- From the Earth we know it takes 1 year for the Earth to orbit the Sun and watching carefully, we find that it takes **Venus 0.615** of our years and **Mars 1.881** of our years to orbit the same star.
- Kepler's Third Law basically states that the square of a Period (P) (the amount of time to orbit another body) is directly proportional to the cube of its semi-major axis distance (A) (here we will use average distance which is a good approximation for orbits that are not too eccentric).
- So our formula looks like this :
- **A^3 is proportional to P^2**
- ***<u>Our question is thus : In terms of Earth's distance from the Sun, how far away are Mars and Venus respectively from the Sun?</u>***
-
- We can make the approximately equal sign into an equal sign by using the Earth as the comparison and setting the problem up this way :
- A^3 (Mars) / P^2 (Mars) = 1 = A^3 (Earth)/ P^2 (Earth)
- So **$A^3 = P^2$**
- (When dealing with units relative to another body, such as the Earth)
- Solve for each of the Planet's Relative Distances in turn.
- We will start with **Mars**.
- We were given the Period of the Planet at 1.881 years. We can only find 1.88 on our D scale which is where we begin.
- Look from 1.88 on D scale to the A scale to find the Period squared. The value falls just short of 3.55 on A scale, so we can say 3.54.
- Now we need to find the cube root of this number.
- The formula states that the square of the period (which we have) equals the cube of the planet's distance.
- We now need to use our K scale as this is the cubing and cube-rooting scale.
- Since the squared value equals the cubed value, we have that, we need to find the cube root of 3.54 from the K scale.
- But first a problem?
- Which one of the 3.54 values, there are 3 of them as you read across the K scale since it goes from 1 to 1 a total of 3 times!?

- ➤ Much like for the A scale where the numbers go from 1 to 10 and then 10 to 100, the same principle holds for the K scale, only here it is 3 times. So it goes from 1 to 10, then from 10 to 100, then from 100 to 1000.
- ➤ So we find 3.54 between the first and second 1. Be sure to estimate the placement answer to 4 place values (first three are significant and the fourth is estimated).
- ➤ Read from it down to the D scale to take the cube root of this value. It aligns with 1.525
- ➤ This is our answer in terms of Earth's distance from the Sun, which means mars is on average 1.525 x farther away than we are.
- ➤ (The given actual answer is 1.524, which places our value in very good company.)
- ➤

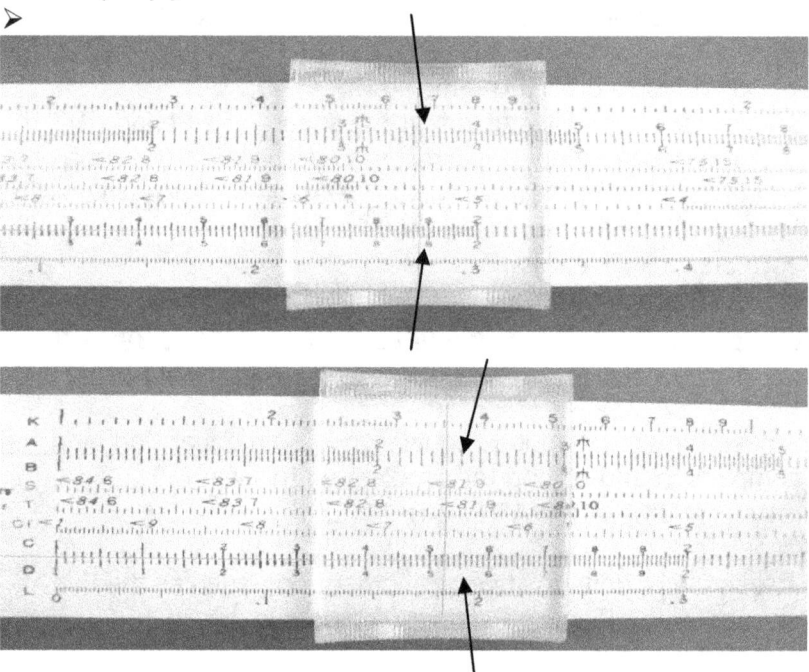

- ➤ Now let's try this with **Venus**.
- ➤ Start on D and disregard the decimal for now. Find 615 and read the square of it on A and come up with 378.
- ➤ But now the decimal. Realize that 6 x 6 is 36 and if we think of the number in scientific notation, 0.615 is **6.15×10^{-1}**. We have to square this number. For the exponents, we add the exponents and come up with -2, so our answer is **37.8×10^{-2}** or 0.378
- ➤ But how to read this on the K scale. Recall I mentioned 1-10, etc, what of decimals?!
- ➤ Think of what it is : 378 thousandths. We can use this fact to look at the scale, so we look at the K scale between the last two ones, or in the last third of the scale. This is the place between 100 and 1000.
- ➤ Find 385 there and read down to D scale once again (see photo) and find the value. 722
- ➤ We read this as Venus having a distance that is 0.722 of Earth's distance to the Sun
- ➤ (the actual standard answer given to this level of accuracy is 0.723).

- ➢
- ➢ LOG-LOG PLOT ACTIVITY
- ➢
- ➢ Given the process of how to do this, now turn to the available table below of the planets known to the ancients and take the log values of both the period and distance for each planet.
- ➢ Treat each as a pair of data points (log of distance, log of period) and graph these on a x-y axis.
- ➢ Draw a best fit line and then determine the slope of this line.
- ➢ You should find that in whole number ratio it should come to the ratio of 2 to 3 since these are the powers of those values.
- ➢
- ➢ *ASIDE*
- ➢ *A further exercise, if you wish, is that the Earth's average distance is 150 million kilometers or 93 million miles.*
- ➢ *Given our results, we can easily compute the numeric distance values for these planets.*

Orbital Data

	Semimajor Axis (10^6 km)	Semimajor Axis (Earth = 1AU)	Orbital Period (d)	Orbital Period (Earth years=1)
Mercury	57.9	0.387	88.0	0.241
Venus	108.2	0.723	225	0.615
Earth	149.6	1.000	365	1.00
Mars	227.9	1.523	687	1.88
Jupiter	778.6	5.205	4332	11.9
Saturn	1434	9.586	10761	29.5

The Efficiency of a Complex Electrical & Mechanical System
Activity #25
Grade Level : High School
Math Level : Calculating

Efficiency is simply the Ratio of Useful Work Output to Work Input. It is a natural consequence stemming from the Laws of Thermodynamics (see associated Rate of Heating Activities). Often in our Activities we have assumed a perfect transference of energy, such as gravitational Potential Energy to Kinetic Energy (such as in the Efficiency Activity), but there is some loss of energy to heat and sound.

This Activity is much like a Rube Goldberg device, though far less sophisticated, where there are a series of steps. We use a light source (perhaps a desk lamp or a flashlight – torch to United Kingdom users, or the Sun itself) to power some solar cells which in turn is connected to a small electric motor which is attached to a pulley with a string attached to a bag of masses (marbles perhaps) to lift them a given distance.

In each case there will be Work Input by some energy source and Work Output. The ratio of Output to Input ideally would be 100%, but in reality, will be less as our results will show. Better still, in combination the final efficiency of our system will be the combined product of each of the steps multiplied together. Test this idea for yourself.

Efficiency is a very critical component of our everyday lives both in real measured amounts and in the processes we engage in – such as when we go out in the car we do several errands and connect the steps in a loop so as not to do multiple trips or re-crossing our path. When it comes to mechanical and electrical processes we find ways to increase efficiency – such as lubricated joints, use of ball bearings in motors, and the like to reduce friction, hence reduce heat therefore wear and tear on systems and convey the energy from one system to another more effectively with less losses. The better we do this, such as turning the stored energy of coal into steam then into electricity that powers our homes and factories, then the less waste there is.

Purpose : To measure the Work Input & Work Output of various subsystems
of a complex system (electrical and/or mechanical) to determine
Efficiency at each step of the system with its energy transformations.

Materials :

- Bag of Marbles,
- Ruler,
- Meter Stick,
- Batteries (if needed for the flashlight – if using a flashlight),
- Known Lumens value Flashlight (the higher the better, 1200+ best) OR,
- Light Bulb (75 W or 100 W) and Desk Lamp (best choice - see Note),
- Solar Cell(s) (the more used the better the system – 3 is good),
- Small Model Electric Motor (1.5 V – 3.0 V, can operate around 80 mA),
- Alligator Clip Wires,
- Pulley that can attach to motor (or one that can be made – see directions for Activity 4),
- Lab Quality Thermometer (Note : instead can use a multimeter with thermocouple),
- Mass Scale,
- Multimeter (possibly with various applications *),
- Tachometer,
- Oven Pad,
- Plastic Crate,
- Styrofoam Sheets (to cover in 2 layers the plastic crate), OR
- Cooler – (Styrofoam is good, small Plastic one with drain plug best),
- Light Socket with Wires and Plug,
- Timer,
- String,
- Tape and Duct Tape,
- Scissors,
- Goggles,
- Stack of Books,
- Rubber Bands,
- Slide Rule

Note : You must have parent permission and supervision in this series
of Activities. Be sure to always act with caution and care. Throughout
the process wear goggles, know how the equipment works, test
materials, be patient, and be willing to redo the tests as needed.

Note : This set of Activities are a series of steps that begins with a light
source (one of the following : the Sun, a lamp and light bulb system,
or a very strong flashlight) that activates a set of solar cells which in
turn are connected to an electric motor which turns a pulley with an
attached string that lifts a small mass (small bag of marbles). In
each step the energy going in (the work in as it is known) is
measured / calculated and the energy out (called the useful work
output) is determined. The ratio of the latter quantity to the former
one is the efficiency of that step of the situation.

Note : This Activity has a number of steps and it may be best to test each individually to see it in operation and then join the subsystems to have a continuous stream of actions. Also this means you may have to run a number of trials to have it operate correctly.

Note : If using the lamp and light bulb system (hence the air calorimeter) be sure that your design does not overheat any surface. Exercise caution and let the system cool after operating for a given time. Do not place anything in contact with the bulb, especially when in operation.

Note : In the case of the desk lamp, it may be best to have a base socket for the bulb (see photo) and a plug that can attach (see photo). This like the other components of the activity require parent consent and supervision. This is best as compared to the desk lamp since it sits low to the ground, hence you may not need stacks of books and the like when using the air calorimeter.

Note : The air calorimeter has two choices : Either construct a make-shift system or use a sufficiently large cooler (the crate can be plastic or Styrofoam). The make-shift one uses a plastic crate (see photo) that has at least one layer of Styrofoam (can have two, one inside and one outside). In any case the key is this : when the light is positioned in the air calorimeter the active bulb must have clearance so as not to overheat any surface. Clearly do not touch the active bulb and let it cool sufficiently before moving. The goal is to have a system that can be 'closed' off and a temperature-measuring system is inserted and read for a short period of time as the air temperature changes (such as with a thermometer or using a thermocouple attached to a multimeter).

Note : * The recommended Multimeter has the conventional applications such as AC & DC measures for voltage, current, as well as for Resistance, but can also have a Photometer for light intensity (and is actually needed if using a strong flashlight as noted above). The model : Mastech MS 8229 is very well suited for this.

Note : If using the lamp and bulb system, you do not need a photometer in your multimeter. The light output will be determined indirectly with an air-calorimeter noted in the directions below.

Photos of Set Up and Other Components :
Note that the last photo of the Air Calorimeter is a work in progress

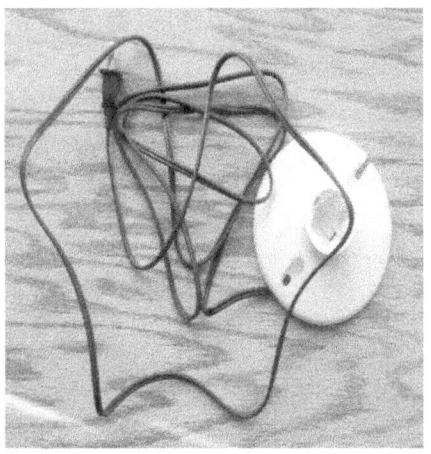

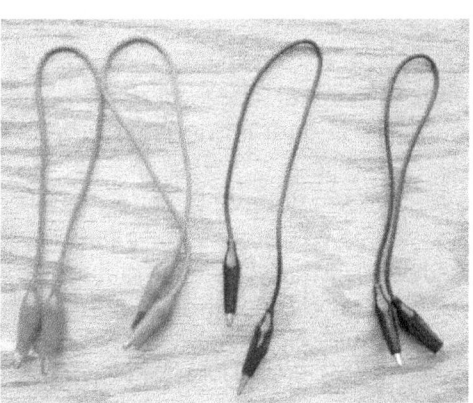

Procedure :

- ### SET UP :
-
- Note : With all of the Activities noted here – Always have parental permission and supervision. It is best to wear goggles. Always employ safe procedures.
- Note : All the Set Up and Measurement Procedures only set up the materials and take measurements, and in each section do the calculation for power in each section then in the Data Analysis section following all of these is where the efficiency for the data is addressed.
- This means that you may have to read the directions in a manner that satisfies what you are doing and where you are at. For example, if you had done Activity 1B (as an example illustration), then there is no need to read 1A nor 1C.
- Next you can then do computations for 1B in the Data Analysis OR continue in sequence for each of the Activities (2, 3, etc) until set up and measurements are completed.
- Note : You may have to do repeated trials and be sure to test each step of the path before having it operate. It is best to look at each component as a step. –
- In the case of multiple trials, it is often best to do a given trial as many as 5 times. Then eliminate both the highest and lowest values and then take the average of the other 3 where this average acts as the intended data measurement value.
-
- NOTE : Your light source determines which Activity 1 (A,B,C) you do :
- Choosing the subset of Activity 1 is determined by how you are able to measure the light source being used :
-
- If you have a Photometer, then you can do any of them,
-
- If you do not have a Photometer, then you must do Activity 1B,
-
- If you use a very strong flashlight, then start with 1A,
-

- If you use a light bulb and lamp, then start with 1B (recommended),
-
- **<u>Set Up of the Air Calorimeter (used only with Activity 1B):</u>**
-
- Note : The initial Air Calorimeter is shown in the set of photos and is the last one. Note that this is not complete, yet.
- What is needed is an open grate crate with thin Styrofoam sheets (can use foam-core poster board instead).
- The Styrofoam is cut so as to line both the inside and the outside of the crate along the plastic.
- It can be held in place by duct tape.
- Note to leave enough to make a 'lid' as well.
- Special Note : If this takes too much time, money, resources, one could instead simply use a Styrofoam Cooler in place of all the aforementioned items!
- An even better choice is a small plastic cooler with a drain plug – the need for a hole is because a thermometer will be inserted to measure the change in air temperature with a lit light bulb inside.
- Returning to the Air Calorimeter Set Up :
- With either the Styrofoam-lined crate or the Styrofoam Cooler, now invert it and pick a corner through which either a Thermometer will be inserted OR where the Thermocouple will be inserted. It is best to make a hole with a thin pencil or better still a knitting needle if available.
- Re-invert the box and use a piece of Styrofoam to construct a partial barrier that can be held in place with tape so that when the Thermometer or Thermocouple is in the Air Calorimeter it is not going to be in direct line of light from the light bulb which will sit opposite the temperature probe slot at the 'top' of the air calorimeter.
- It can be seen that there is a reference to the 'top' in a quotation-mark manner. This is because ultimately the box will be inverted and placed atop the 'lid' which lies flat on the ground and on top of the 'lid' in the center of it is the light bulb in a small light fixture (see photos) and connected with an appropriate electrical wire system so that it can be plugged in.
- The first main measurements to make here are the Length (L), Width (W), and Height (H) of the interior of the Air Calorimeter space. Record these measurements. Measure these to the nearest $1/10^{th}$ of a centimeter.
- This completes the Air Calorimeter Set Up and initial measurements –
- The Air Calorimeter needs the Volume (V) calculated from it and when coupled with the Density of Air (ρ_A) will yield the Mass of air in the Air Calorimeter (m_{air}) (do this calculation).
- Use this Air Calorimeter calculation with Activity 1B next
-
- If you use the Sun, then start with 1C,
-
- **<u>The goal of Activity 1 is to measure the power of the light source, which will be the Input Work for Activity 2</u>**
-
- **<u>Activity 1A : A Strong Flashlight as the Light Source</u>**

-
- Note : In order to use a flashlight, it must be a very luminous type, such as 1200 lumens or more. These can be expensive, so it is recommended to do Activity 1B or 1C.
- Note : If using a flashlight, it is best to only use an LED type and not other luminous source lamps which can be very hot. The flashlights described here are strictly battery operated and LED type and this is the only type recommended.
- Note : Since most, if not all very luminous lamps are LED types, it would be very difficult at best to use a simple air calorimeter, so the best measurement system would be to use a photometer to determine lux output of the light and then to translate this into Power.
- Note : The fastest way to test whether or not you can use this path is to quickly test your solar cells – connect them to the motor and see if this type of light can power the system. If it cannot, then you need to try either another method or find still a brighter light (if accessible – note that the more luminous types cost more)
- The first thing to measure is the Power Input into the light source.
- Use the Multimeter to measure the batteries both for Voltage (V_{1A}) and Current (I_{1A}) when not connected to the lamp. Be sure to know how to operate the multimeter to take these measurements. It is best to have the batteries connected in the same manner as they are in the flashlight (in series essentially) and measure them operating together. Record these results. The results from these measurements will be calculated and are considered Work Input #1.
- Now to measure the flashlight output and translate the light output into Work Output #1 (which in turn acts as Work Input #2).
- The most effective measurement of measuring the light is to use a photometer.
- Have the flashlight positioned so that it has an effective cone of light and the multimeter photometer sensor will be centered facing it at 1.0 m distance.
- The key to the measurement is to have as little outside light as possible, so you must find a means to read the meter (most digital have illuminated screens) such as drawn shades, etc.
- The reason for 1.0 m distance is the fact that the lux (lx) meter will be measuring lumens/meter-squared (lm/m^2)
- Record this lux measurement (L_{1A})
- Move on to Activity 2
-
- **Activity 1B : A regular Electric Lamp as the Light Source**
-
- With the Set Up completed first for the Air Calorimeter (see above) now we can measure indirectly the Power Input and Power Output of the Light Bulb as a Light source in our Activity
- Note as with any bulb, do not touch it when active nor handle it for some time after it is has been active since it is hot. Do not place objects on or too near the active light bulb. Be sure to have parental permission and supervision.
- First measurements are to determine the power going into the lamp.
- We can simply take the measure of the light bulb, say 75 W (a good recommendation), as the measure of the input power to the system.

- This value will be considered the Work Input #1 (W_I) and can be written on the table. If you are not measuring the light bulb's input power directly (requiring a specialized multimeter) now move on to the light output measure as work output #2.
- Light output as Work Output #2 Measurement :
- The output power can be measured either directly, using a Photometer as we did in the flashlight case (only here we use a light bulb), so if this is your method, read the directions in section 1A and replace flash light with the light bulb and do the same thing.
- The other measurement method is indirectly by measuring the heat the light generates and use this to subtract from the input power to find how much useful power is turned into light.
- Side Note : It may be a good experiment on its own to try both methods to see how similar or different these values are and consider why they are different!
- Indirectly measuring Power Output of the Light Bulb using the Air Calorimeter :
- Note that the Air Calorimeter must be completed (see directions above) for this next section.
- Here the bulb is in the socket and the Air Calorimeter is placed over it.
- Initially do not activate it.
- Use either a lab quality Thermometer inserted through the hole in the top (note that it needs to have a barrier to block the direct light upon the thermometer, such as a piece of white poster board or Styrofoam) or you can use a thermocouple attached to a multimeter for temperature readings.
- Take an initial temperature reading with the light off (T_i).
- Now start the timer, and take temperature readings every 2 minutes for at least 10-16 minutes total time (t).
- At this point you have enough measurements to be able to compute the amount of energy given off as heat by the light and then with the time   (and proper conversion factors) convert this into watts.
- This is done by creating a graph of Temperature (y-axis) versus Time (x-axis), draw a best fit line and determine slope.
- With the slope you can now determine the heat energy of the system by using the specific heat capacity of air and the volume of air with this slope value – determine this power value.
- Note : This is NOT the power output of the light that we are going to use. We need to take this value and subtract from the Power Input value for the light and the difference is the power output of the light! This difference is the power value for the light we are going to use.
-
- **<u>Activity 1C : The Sun as the Light Source</u>**
-
- We have two choices here :
- Either make an assumption as to the power output of the Sun OR
- We can determine the value of the power output of the Sun by doing the Solar Constant Activity noted in the book.
- Note that the Efficiency in Activity 1 cannot be computed since the input is considered to be the same as the output (though not technically true, one could do the computations of how much of the Sun's energy is given off as light as compared to the amount of energy generated by fusion in the core, but that is another matter)

-
- Regardless of which of Activity 1 (A,B,C) you have done, now move on to the rest of the Activities in sequence :
-
- **Activity 2 : The Light Source acting on the Solar Cell**
-
- Regardless of which light source you have used, you now have a Power or Work Output #2 for a given light source to examine in Activity 2.
- This value will now act as Power or Work Input #2 for Activity 2.
- That logic will continue for each of the Activities, where the Work Output from the prior step becomes the Work Input for the next step.
- When it comes to Efficiency for a given step, it becomes the ratio of the Work Output for that step over the Work Input to that step.
-
- Now we are going to examine Power Output#2 for Activity 2, which comes from the use of one or more Solar Cells.
- Note : It is best to use 3 solar cells rated at 1.5 V and connected in series with each other for later power needs of the electric motor.
- With the Solar Cells connected properly and receiving the light from the light source, now use your multimeter to do the following :
- First, with proper settings on the DC Voltage, and connections of the cells to the meter (it is connected in parallel with the cells) measure the voltage of your solar cells with the light acting on them.
- Record this voltage value (V_{sc})
- Second, now reconnect properly the multimeter so as to measure DC Current (it will be in series with the cells) and be sure to have it on the proper setting so that the circuit does not blow – it is best to begin at the highest setting and then turn down to incrementally lower settings and noting the values along the way.
- Record the current value (I_{sc})
- You can readily calculate Power Output#2 here and do so, but can easily continue to the next Activity (3) where this acts as Power Input#3
-
- **Activity 3 : The Solar Cell acting on the Electric Motor**
-
- From Activity 2 we now have the Power Input#3 (which is Power Output#2) for this Activity.
- Now connect the Solar Cells to your Electric Motor.
- Test it to make sure it operates correctly
- Note, it is best to have a stand for the Electric Motor. A good idea is a stack of heavy books where a ruler is sandwiched somewhere near the top of the stack and jutting out a couple of inches. The Electric Motor is then held in place by a rubber band or two on the motor. Be sure that as little vibration is going on as possible when it is in operation. Also do not run the motor too long in any of the Activities so as to prevent overheating.
- Once the motor is operational connect small cut-out thin poster board circular disk (about ¾" diameter) to the axle of the motor so that it rotates.
- Measure and record the diameter of this disk in centimeters (d)

- Use the black reflective tape that comes with your Tachometer and attach a small piece to one edge of the disk –
- And to promote balance, attach a small piece of masking tape of approximately the same size and mass to the other edge of the disk so that it rotates in a balanced fashion
- Before using the disk with the tape on it, you need to measure its mass (m_d) and record this value in the table.
- Let the motor operate and use the Tachometer to measure the angular speed of the system. (ω)
- Note : We are going to assume that the motor has 'no load' on it, though there is one, hence this will affect the outcome slightly. In our next set of measures we are placing a significant load on the motor, though.
- Now Use the mass, diameter, and angular speed to determine the rotational kinetic energy of the system (KE_R)
- In the calculation section we are going to use this calculation for rotational kinetic energy as the power of the system to determine efficiency.
- Note : We are then going to make some critical assumptions :
- We will assume that the rotational kinetic energy started at zero
- We will assume that in each moment of time the rotational kinetic energy is constant
- With the value of rotational kinetic energy determined, we can then say that the system has this output of energy per unit time (1 second) and hence this same number of watts as the Power Output#3
-
- **Activity 4 : The Electric Motor acting on the Load**
-
- The Power Input#4 is the same as the Power Output#3 from above
- Now to the Electric Motor attach a pulley system (solid rotating disk – which may be purchased, but if unavailable it is best to use poster board cut into 2 disks approx. 1-2" diameter which are glued onto a central foam-core poster board disk of a smaller diameter (½" to 1.5" depending on the outer disks) leaving at least ½" to the outer rims of the make-shift pulley) – On the pulley you make as well you might want to cover the foam-core area with a thin piece of masking tape.
- Next have a piece of string less than 1.0 m in length but it should be longer than 50 cm.
- The length will be affected by how far in the air the motor is based upon the use of the stack of books as noted in the prior Activity.
- Test to be sure the pulley system is operational
- With an operational system we need to do a few measurements :
- Measure the diameter of the inner foam-core disk (ds) and record this.
- Measure the mass of the entire pulley disk system (mp) and record it.
- Now attach the string to the pulley and place a small mass on it (ideally use a plastic sandwich bag with a marble or two)
- Again test the system
- With further tests continue to add masses (marbles) until the system can still lift it but it requires some effort.
- Note : You may want to have back-up pulleys and be sure to keep the tension on your pulley system (a piece of masking tape may help) if needed

- When the necessary mass is in place, measure the entire mass (M) on the scale and record this value for the load of the system.
- Before activating the motor, be sure to measure the distance it will travel from the ground to some given height (H) and record this value.
- With all in place, now Use a stopwatch and measure the amount of time (t) for a given trial for lifting the mass.
- Perform this measurement 3 times and average the time (t_{ave}) which is used.
- From the measurements, first :
- Calculate the new rotational kinetic energy of the motor using the pulley system by turning the linear kinetic energy of the system into rotational kinetic energy.
- The linear kinetic energy is determined from the distance traveled by the bag of marbles divided by the amount of time it takes for them to travel.
- The gravitational potential energy of the system is determined by the mass of the bag of marbles times the acceleration due to gravity times the distance (height) traveled.
- Here the change of gravitational potential energy (PE_g) over the amount of time (t_{ave}) as a power along with an additional factor of the rotational kinetic energy treated as a power of the new disk system will be used in combination to determine the Power Output#4 in watts for our system. (see calculation section for this efficiency calculation)
-
- **Analyzing the Data in All Activities :**
-
- Note : Each of the Sections will first be analyzed for their Power Input and Power Outputs as needed – this will be followed by the Efficiency Calculations for each of the Activities respectively since these are independent and require the Output Work from the prior Activity (which now acts as the Input to the Activity in question) and the Output in a given step is the measurement from that Activity :
-
- **Efficiency Calculations :**
-
- Initial Efficiency of Activity 1 :
- Note that this cannot be done if using the Sun
- For all other calculations one merely takes the ratio of the Work Output, which is the energy of the light source over the Work Input to the light source (such as the battery power or wattage of the bulb)
-
- Efficiency of Activity 1-2 :
-
- This is the ratio of the Work Output as determined from the Power of the Solar Cells as determined from the calculation of Power from the measurements of Voltage and Current to the Work Input which is the Work Output from the last Activity which is the Power of the light source
-
- Note : Independent of which light source used, you have determined its power which acted as initial work out #1 and now it will be used as Work In #2.
- In this calculation, use the readings from the Solar Cell (Voltage and Current) and determine Power from this. This Power will be used as Work Out #2
- Now use Work Out #2 & Work In #2 to determine Efficiency for this step of the Activity

-
- Efficiency of Activity 2-3 :
-
- This is the ratio of Work Output, which is the Rotational Kinetic Energy of the Motor being treated as Work to the Work Input of the Solar Cells to the Motor
-
- Efficiency of Activity 3-4 :
-
- This is the ratio of the combination (addition) of the gravitational potential energy of the bag of marbles lifted plus the rotational kinetic energy of the pulley disks of the motor to the input energy of the motor which is the initial rotational kinetic energy of the motor as determined in Activity 3
-

Data :

Air Calorimeter for Activity 1B :

Dimension	Measure (cm)
L : Length	
W : Width	
H : Heigth	
V : Volume	cm^3

Mass of Air in Air Calorimeter (m_{air}) : _____ g
    Determined from air density and calorimeter volume

Activity 1A :

Flashlight :
Voltage (V_{1A}) : _____ V
Current (I_{1A}) : _____ A
Power (W_I) : _____ Watts
    Power from calculation and is Work Input#1

Distance of photometer : 1.0 m
Photometer Measure (L_{1A}) : _____ lux
Calculated Work Output #1 : _____ watts
    Note that this is Work Input #2
    This measurement is also for the Lamp if there is no Air Calorimeter

Activity 1B :

Noted Lamp Wattage (W_I) : _____ watts
    This can be Work Input #1

Calculated Work Input #1 (W_I) : _____ watts

Use of Air Calorimeter :

Time (min)	Temperature (°C)
0	T_i
2	
4	
10 – 16	T_f

Change of Temperature (ΔT) : _____ °C
Quantity of Heat added to Air Calorimeter : _____ J

Calculated Slope of Line ($\Delta Temp / \Delta time$) : _____

Calculated Power Loss due to Heat in Air Calorimeter : _____ watts

Net Work Output of Light (W_{O1}) : _____ watts
    Calculated from $P_{Net} = P_{max\ input} - P_{loss}$
    Note that this is Work Input #2

Efficiency for Activity 1, 2, 3, 4 :

Note : To change this to all other Activities, simply change the 1 to 2
    and so on as needed for each of the Activities (2, 3, 4)
    This means you need to do this calculation for each Activity where
    the proceeding Activity is the Input Work and the Work measured
    in that Activity is the Output Work

$$\text{Efficiency} = \frac{\text{Work Output \#1, then 2, et al}}{\text{Work Input \#1, then 2, et al}}$$

Activity 1C :

Use Solar Constant or Do Solar Constant Activity
This is Work Output #1 and is Work Input #2

Activity 2 :

From Activity 1, we now have Work Input #2

Solar Cells :

Total Voltage (V_{SC}) : _____ V
Total Current (I_{SC}) : _____ A

Total Work Output#2 (W_{O2}) : _____ watts
      Acts as Work Input#3 (W_{I3})

Activity 3 :

Disk diameter (d) : _____ cm
Disk radius (r) :   _____ cm
Disk mass (m_d) :   _____ g

Angular Speed of Motor (ω) : _____ rev/s

Rotational Kinetic Energy of Motor (KE_R) : _____ J
      Note : This is the assumed Power - Work Output #3
      Note : This is the Work Input #4

Activity 4 :

Inner Disk :
Disk diameter (d) : _____ cm
Disk radius (r) :   _____ cm
Disk total mass (m_T) :   _____ g

Circumference of Disk (C) : _____ m

Length or Distance Traveled (H) : _____ m

Mass of Marbles used (M) : _____ kg

Potential Energy of System (PEg) : _____ J

Time to travel distance H (t) : _____ s

Calculated Linear Speed (v) : ($\frac{H}{t}$) : _____ m/s

Calculated Rotational Speed (ω) : _____ rev/s

Rotational Kinetic Energy (KE_R) : _____ J

Total Energy of System (E) : _____ J

Total Power of System (Work Output #4) : ($\frac{E}{t}$) : _____ watts

Calculations :

Be sure to use your Slide Rule!

Efficiency :

$$Eff = \frac{\text{useful Energy Out}}{\text{Energy In}}$$

$$Eff = \frac{W_{out}}{W_{in}} \qquad (\text{ W = Work })$$

Power :

$$P = \frac{W}{t} = \frac{\Delta E}{t} \quad (\text{ General Formula : W=Work, } \Delta E = \text{ Change of Energy, t=time })$$

$P = V*I$ (Electrical : V=voltage, I=Current)

$P = N*S$ (N = No. of Lux or L_{1A} and S is number of square meters)

$P_{Net} = P_{max\ input} - P_{loss}$ (Determines Net Power in Light Output)

Energy :

$W = F*d$ (General Formula for Work : W=Work, F=Force, d=distance)

$W = \Delta E$ (General Formula : W=Work, ΔE = Change of Energy)

$PE_g = m*g*h$ (Gravitational Potential Energy : m=mass, g=acceleration due to gravity, h=height)

$PE_{el} = \frac{1}{2}*k*x^2$ (Elastic Potential Energy : k=spring constant, x=displacement of spring)

$KE = \frac{1}{2}*m*v^2$ (Kinetic Energy : m=mass, v=velocity)

238

$KE_r = \frac{1}{2}*I*\omega^2$ (Rotational Kinetic Energy = Moment of Inertia *
                                    Angular Speed)

$Q = m*c*\Delta T$ (Heat Needed-Emitted = Mass * Specific Heat Capacity
                       of a Material * Change of Temperature)

Other Relations Needed :

$\Delta N = N_f - N_i$ (The change of a given variable is the difference of a
                          final value and an initial value)

$m = \frac{\Delta Y}{\Delta X}$ (Slope for a given line on a graph)

Area of a Square : $A = L*W$
Volume of a Rectangular Solid : $V = L*W*H$

$C = 2*\pi*r$ (Circumference from the radius of a circle)

$r = \frac{d}{2}$ (radius is half the diameter of a circle)

$F = m*a$ (Net Force = mass * net acceleration)

$F = k*x$ (Hooke's Law : Force = Spring Constant*Distance Stretched)

Speed :

$v = \frac{\Delta d}{\Delta t}$ (Linear Speed = Distance Traveled / Elapsed Time)

$\omega = \frac{v}{r}$ (Angular Speed = Linear Speed / Radial Distance)

Moment of Inertia Formulae :

$I_{disc} = \frac{1}{2}*m*r^2$ (m = mass, r = radius)

$I_{rod-mid} = \frac{1}{12}*m*L^2$ (L = length)

$I_{rod-end} = \frac{1}{3}*m*L^2$

<u>Constants :</u>

1 m = 100 cm
1 kg = 1000 g
1 lux = 1 lm/m^2
1 kg = 2.2 lbs
1 in = 2.54 cm
1 ft = 12 in
1 lb = 16 oz.
1 oz. = 28.3 g
1 cal = 4.184 J
1 Watt = 1 J/s
1 kW = 3,414 BTU/hr

Acceleration due to Gravity : $g = 9.8$ m/s^2 = 980 m/s^2 = 32.4 ft/s^2

Specific Heat Capacity of Water
$$c = 1.0 \frac{cal}{g*°C} = 4.186 \frac{J}{g*°C}$$
Specific Heat Capacity of Air (at room temp of 25° C)
$$c = 0.242 \frac{cal}{g*°K} = 1.012 \frac{J}{g*°K}$$

Actual Solar Constant : 1.366 kW/m^2
Corrected Amount received at Earth for Solar Constant :
342 W/m^2

1 kW = 10^3 watts
1 kW = 3,414 BTU/hr
1 BTU/hr = 2.93 x 10^{-4} kW

Constant Used to Convert Light Into Power
1 lux = 1.46 x 10^{-7} Watts/sq.cm. (at 555 nm (green light))

1 lux = 1 x 10^{-4} lumen/sq.cm.
1 lux = 0.0929 lum/sq.ft. (also ft-candles)

<u>Conclusion :</u>

One of the first set of questions to consider are these : What are the
types of energy involved in your system? What type of energy
transformations take place in this system? When it comes to efficiency,
what becomes of the energy that seems to be 'lost' when going from one
step to the next? How do you account for it (assuming conservation of
energy applies)?

What do your results show for Efficiency for component parts? How do you think this relates to the whole system? Why do you think some subsystems were more efficient than others? Would do you think can be done (and perhaps tested) to change or effect efficiency of the whole system or some subcomponent of it? If the system were less complicated, would efficiency be a larger value or not (bearing in mind, would it depend on the components involved)?

Alternative Tests to consider for comparison (with parental permission and supervision) :

1) Try a different wattage bulb
2) Try a different type of bulb (fluorescent)
3) Try a different distance for a given bulb intensity
4) Try a different size solar cell, or more solar cells
5) Try a different engine size
6) Try a different number of marbles as the load to be lifted
7) Try a different initial energy input of your design and testing (with parental permission and supervision)

Activity #26
Solar Cell Examinations
Grade Level : High School
Math Level : Calculating

Solar Cell Measures Activity :

This Activity has the following Prelude on Solar Energy and Photovoltaic Cells for those interested in the subject. Otherwise you can move ahead to the Activity that follows concerning an investigation of these photovoltaic cells.

The Sun provides an enormous amount of energy for the Earth. The energy is discussed with more details and numbers in the Solar Constant Activity #, but it can be said that the Sun's energy reaching the Earth's upper atmosphere is on the order of 174 petawatts of power (peta meaning 10^{15}). This energy drives the wind, the waves, provides heat, light, and energy for life forms of the planet.

Solar Energy in many forms, especially in the present world of technology, is used for many reasons. The most obvious is light, which is utilized for everyday visual uses. The most obvious employment of the Sun outside of regular light and heat is the growing of crops by humans which takes advantage of photosynthesis by plants which converts carbon dioxide (CO_2) and water (H_2O) into molecules to provide energy, namely sugar. The heat of the Sun will be discussed more thoroughly in the Heating Activities # (if employing the Sun), both in human application and its appearance in nature, such as weather.

Humans have basically utilized light in two ways categorically : passive and active solar energy technologies. These classifications result from the way the sun's energy is captured, converted, and distributed. The Passive Solar Energy Technology will be explored in more detail in the Rate of Cooling Activity #28 indirectly, but it can be said here that it is based upon the thermal properties of materials positioned in favorable ways to take advantage of the Sun' energy, heat, and light.

Active Solar Energy ideas are ones that take the sunlight and convert it into other forms of energy (the most common being electricity) for use. In this technology there are two chief types : 1) concentrating solar power - focused sunlight and the use of Stirling engines or for boiling water to use with steam turbines 2) utilizing the photoelectric effect - photovoltaics cell use (which is the focus of this Activity).

In concentrating solar power, here optics such as lenses or parabolic mirrors are used to focus the sun's light energy onto a region or point to act as a heat source to actively power a power station, such as a Stirling engine. These heating systems often heat a fluid of some form to cause it to do work, which can then activate an electric generator. Others can be used to boil water to turn turbines as well.

The other type of Active Solar Technology are Solar Cells –aka Photovoltaic cells (the latter term can apply to any and all light whereas the former applies to sunlight) which convert sunlight directly into electricity through the photoelectric effect.

The term 'photovoltaic' comes from 'photo' Greek for 'light' and 'voltaic' derived from the Italian physicist Volta and the term 'volt' came from which is the unit of electro-motive force, so 'voltaic' is intended to mean 'electric'. This idea was first written about but not constructed in 1839 by the French physicist A. E. Becquerel. The first created one came in 1883 by Charles Fritts, who used coated the semiconductor selenium with a thin layer of gold at the junctions. This first one was on 1% efficient. Not long after the first solar cell based on Heinrich Hertz 1887 photoelectric effect came from the Russian physicist Aleksandr Stoletov. Even Albert Einstein played a part in the history of the solar cell with his 1905 paper explaining the photoelectric effect for which he won a Nobel Peace Prize in 1921. In time more research and modifications were made to the solar cell. A much more efficient form using a diffused silicon p-n junction was developed by Chapin, Fuller, and Pearson in 1954. Some solar cells today have reached values of 24%, and even 42% absorption efficiency (the latter one with concentrated light), well above the average in the industry which is at 12-18%. (Note that this stands in contrast to sunlight-to-electricity efficient, which is lower). Today there are even Megawatt solar power generating plants being built.

This type of electrical energy production is not only found in small devices, such as arrays for charging batteries and uses in calculators, but is always in use in space (satellite and space station power sources) and it is finding a growing market worldwide here on the ground. The efficiency of them is increasing, their fragility is decreasing, and their costs are coming down. There are even bendable solar cell arrays that are used as shingles on roofs these days. As of

2010 power generated by them is going on in over 100 countries. It is estimated that 4800 GW could be harnessed this way yet only some 21 GW are being used at present. Since 2002, photovoltaic production has risen by 20% per year making it the fastest-growing energy technology. Historically the first commercial use of it was in 1966 on Ogami Island in Japan to change the Ogami Lighthouse from gas torch to fully-sufficent electrical power. The three leading countries in use of this technology are Japan, Germany, and the United States. Presently most stations range from 10-60 MW capacities and in the near future 150 MW or more are projected. All of these worldwide efforts are driven by the need for renewable energy sources and their utilization.

Solar Cells are made of semiconducting materials (these are elements that are in the metalloid region of the Periodic Table between metals and nonmetals). Over the cell is a clear covering of glass or plastic to allow light in, but to protect the semiconducting wafers. These cells are encapsulated and care connected in series and/or parallel connections depending on the need for greater increases in voltage and/or current and for the overall power of the system. The arrangement is called an array.

The solar cell generates a DC (direct current). The power output is measured in watts or kilowatts (depending on the number of cells in the array). To determine the number of cells needed and the arrangement of cells needed, the energy needs calculated in watt-hours, kilowatt-hours, or even a energy per day as kilowatt-hours per day are often used. A quick mental rule is this : Average Power is equal to 20% of Peak Power.

The power from a solar cell array can be fed directly into the electric power grid through inverters. If these arrays are stand-alone the energy is often stored in rechargeable batteries if it is not immediately being used (such as powering a device). Smaller panels can be used as chargers and direct power sources for cellular phone chargers, solar-powered calculators, solar lights for things like bikes and solar-charged camping lanterns.

How do they work? When photons encounter the surface, some pass right through (since their energy has no effect on the materials), some reflect off the surface of the cell, while some photons have the right amount of energy that corresponds to the 'silicon band gap'.

These photons that hit the solar panel and are absorbed by the semiconducting materials energize the electrons in the valence band. These electrons are regularly tightly bound in covalent bonds between neighboring atoms. The energy they are given excites them into the conduction band where it can move more freely in the semiconductor. Where the electron was, now has what is called a 'hole'. This creates a potential for electrons in surrounding molecules to move into it. Hence electrons move one way while the holes move in the more-or-less opposite direction. Next these negatively-charged electrons are knocked loose from their atoms are allowed them to flow though the semiconducting material and produce electricity. The cells composition only allows for the electrons to move in one direction. Hence, when cells are connected to each other will have an overall voltage and current flow and produce a measurable and useable amount of DC electricity.

The most commonly known solar cell configuration is called the p-n junction and is made from silicon laced with other semiconducting materials. Basically it is a materials where a layer of n-type silicon is placed in direct contact with a layer of p-type silicon. That is just a mental model, while typically the layer of silicon has one side diffused with a n-type dopant into a p-type wafer (or vice versa) so that the effect is the same.

In the p-n junction type there is a diffusion of electrons from the region of high electron concentration (the n-type side of the junction) into the region of low electron concentration (the p-type side of the junction). When the electrons diffuse across the p-n junction, they recombine with holes on the p-type side. This does not continue indefinitely, since a charge build up results in an electric field, which in turn, creates a diode that promotes charge flow (aka drift current) which opposed and eventually balances out the diffusion of electrons and holes. The area that no longer contains any mobile charge carriers is called the space charge region.

The solar cell is connected to an external load at its positive and negative points. The voltage measured is equal to the difference in the electrons in the p-type portion and the holes in the n-type portion. With voltage, when connected to any load (which has resistance) then a current can occur (Ohm's Law).

The amount of Power from a solar cell easily can be seen from the amount of light falling on it, which is one of our Activities. Also, the number of solar cells used and how they are arranged will affect the amount of power available from a solar array. The two common types of arrangements are called : Series & Parallel arrangements.

Like in Ohm's Law where Resistors are connected one after another and hence are a Series, the same is true for solar cells. Here the positive lead connects to the next solar cell's negative lead and so on. What is left is the first one has a negative lead not connected, and the final one has a positive lead not connected. These can be connected to a device or a multimeter for determining its readings. One should find that the voltages should sum up and hence increase, while the current values should remain constant.

In the Parallel arrangement, the positive leads are all connected as are the negatives to each other in two lines across all of the solar cells. From the lead or final solar cell to additional lines are connected for connection to a device or a multimeter. Here all the voltages are the same, so it will not increase, while the currents will all sum up and have a larger value.

One could also be more creative, possibly wanting both greater voltage and current and create, for example, two separate lines where each line has two solar cells in series with each other while the two lines are in parallel to each other. This system will add up in a way so that it has both increases in voltage and in current values.

Our investigations will take the solar cell and concentrate on constant light sources at different distances, angles, and even the amount and/or type of light reaching the solar cell. The last activities examines the connections of more than one solar cell and looks at series and parallel arrangements.

One idea not explored here is distance of the light source from the solar cell, since this addressed in the Inverse-square Law of Light Activity. Also the solar cell can be used in the Solar Constant Activity #45.

The Solar Cell Investigations :

      I.   **Solar Cell Power due to Exposed Area of Cell**
      II.   **Solar Cell Power due to Angle of Light Exposure**
     III.   **Solar Cell Power due to Light Intensity**
     IV.   **Solar Cell Power due to Wavelength of Light**
      V.   **Solar Cell Power when connected in Series**
     VI.   **Solar Cell Power when connected in Parallel**

Activity I : Solar Cell Power due to Exposed Area of Cell
Purpose : To measure and determine the effect on voltage and current measurements (to calculate power) for a constant light source at a given distance while changing the amount of solar cell surface area exposed to the light.

Activity II : Solar Cell Power due to Angle of Light Exposure
Purpose : To measure and determine the effect on voltage and current measurements (to calculate power) for a constant light source at a given distance while changing the angle that the light strikes the solar cell.

Activity III : <u>Solar Cell Power due to Light Intensity</u>
Purpose : To measure and determine the effect on voltage and current measurement (to calculate power) for a constant distance light source of varying intensity light emitted and received by the solar cell.

Activity IV : <u>Solar Cell Power due to Wavelength of Light</u>
Purpose : To measure and determine the effect on voltage and current measurements (to calculate power) for a light source emission filtered by color filters to allow certain wavelength access to the solar cell.

Activity V : <u>Solar Cell Power when connected in Series</u>
Purpose : To measure and determine the effect on voltage and current measurements (to calculate power) when a solar cell array has its cells connected in a series circuit manner when receiving a varying light source intensity.

Activity VI : <u>Solar Cell Power when connected in Parallel</u>
Purpose : To measure and determine the effect on voltage and current measurements (to calculate power) when a solar cell array has its cells connected in a parallel circuit manner when receiving a varying light source intensity.

Materials :

- Three 1.5 V Solar Cells (only one if not doing series or parallel),
- 6 Alligator Clip wires or wires to connect the solar cells (2 for one),
- Multimeter,
- Bulbs : 25 W, 40W, 60W, 75W, 100W,
- 2 Small Lamps (regular post type and flexible head),
- Protractor,
- Meter Stick or Measuring Tape,
- Color Filters (R, G, B),
- Piece of Dark construction paper or cardboard,
- Scissors,
- Tape,
- Graph Paper,
- Small penlight to read and write by,
- Item(s) to create angle for solar cell to rest on and be at – can be a stack of books/magazines, use of poster board or cardboard folded, or some other imaginative item to get the job done,
- Slide Rule

Procedure :

1) All Activities use a Light Source (Lamp with at least one bulb used), the multimeter, and at least one solar cell, and all have their calculations done with the slide rule.
2) Recommended : Use rectangular, 1.5 V solar cell
3) Note that since the effect of the light on the solar cell is the primary measure of interest, try to do most of the measurements with as little external light as possible. (have the small light available to write things down).
4) Exercise safety in using the lamp : Do not touch the bulb, Leave it unplugged until the bulbs are properly in place and then activate it, do not place items near or on the bulb (such as the solar cell or multimeter).
5) It is best between trials to turn off the light, turn off the multimeter, and then set it up to operate for that given trial.
6) Recognize that not only do you turn the dial (typically) to change from reading voltage to reading current on the multimeter, you must also change the configuration of the test wires. Be sure of the instructions on how to properly operate the multimeter for voltage and current readings.
7) Since the solar cell is part of every experiment, always have the positive and negative leads coming from one (or more) the solar cell(s). If properly done, there are only two wires, one positive and one negative to touch or connect to with the multimeter leads.
8) Note : Lamps are used so as to control the amount of light, its distance, angle, et al. But once there is an understanding of the results in using a solar cell, one could use their intended source, namely the Sun on sunny days and try these experiments using it when and where possible. Be sure to not look directly into the Sun. Use shadows of pencils act as guides for pointing items directly at the Sun. Also employ safety when using bulbs as well – do not touch active bulbs nor even when turned off – do not look directly into the light. Employ common sense safety procedures.
9)
10) **Activity I : Solar Cell Power due to Exposed Area of Cell**
11) Measure the Length and Width of the Solar Cell. Calculate its Area.
12) Use the dark construction paper. Measure and draw a rectangle 3 times the width by one length of the solar cell dimensions.
13) Cut the first piece to be the length and width of the solar cell.
14) For the second, cut a rectangle the width and for the length measure 75% of the length and cut it there.
15) Save the 25% piece too since this is the final piece.
16) For the third, cut a rectangle the width and for the length measure 50% of the length and cut it there.
17) Each of the rectangles are used to cover the exposed area of the solar cell in turn as noted in the data table.
18) For this Activity, it is best to use the flexible light since it can be directed onto the solar cell lying flat on the table or floor. If not, just be sure not to move the lamp once in place so that the amount of light available is the same for each trial.
19) In setting up the solar cell, be sure to connect the wires so that you can connect a meter to take readings.

20) Choose a Wattage bulb (60 W recommended) and a distance for the light (about 25 cm recommended). Plug in and activate it.

21) Choose an order (like the one noted in the data table) for percent exposed.

22) Start with the Voltage readings [V]. (note readings might not be in volts (V), depending on distance and bulb intensity)

23) After each reading, record it, turn off the light and place the appropriate cover on the solar cell for the next reading.

24) Reactivate the light and continue the readings.

25) Next configure the multimeter to read Current [I]. Check to see it has the proper setting (note that readings may not be in amps (A) so be sure to correct for this).

26) Note : 1000 mX = 1 X (X can be volts or amps)

27) Go through the process and record all of the current values.

28) Calculate the Power using your slide rule.

29) Create 2 graphs of this : First a Bar graph of Power versus Percent Exposed. Second a Line graph of the same variables and compute the slope of the best fit line for this case.

30) A good follow up to this activity is to try other wattage bulbs to see if the results vary or not.

31)

32)   Activity II : Solar Cell Power due to Angle of Light Exposure

33) In this Activity, use the lamp (60 W bulb recommended), the solar cell, the multimeter, a stack of books and the protractor.

34) The best set up is using a standard post lamp set at a constant distance (25 cm to 50 cm).

35) The solar cell is atop a duel set of books and goes from lying horizontally to standing vertically, propped up by paper/magazines/cardboard.

36) The key is to have the center of the bulb and the solar cell when standing up directly in line (this is referred to as vertical or 90°)

37) The angle that the solar cell is at is measured (see data table below)

38) The angle is with respect to the ground and is the compliment of the angle with respect to the solar cell. Determine this.

39) At each angle, measure the Voltage and the Current using the multimeter.

40) When done, calculate the Power of the solar cell for each of the angles.

41) Create both a Bar Graph and a Line Graph of Power versus the Angle and compute the slope of the best fit line for the line graph.

42) Like the other Activities, try other wattage bulbs for comparison.

43)

44)   Activity III :

45) For this Activity, use the standard post-style lamp, have all the bulbs available, and use the meter stick to determine distance between the lamp and the solar cell.

46) As in Activity 2, you could have the solar cell stand vertically and facing the bulb (though no matter its orientation to the light if held constant it will be okay).

47) Choose a distance between the lamp and the solar cell that is held constant (recommended 25 cm to 50 cm).

48) Set up the lamp with the lowest wattage bulb first and work your way through to the highest wattage bulb for each of the trials.

49) With a given bulb in and on measure both the voltage and current produced in the solar cell and record your results in the table.
50) When done with the data table, calculate the Power for each of the bulbs.
51) Graph in a Bar Graph and a Line Graph the results of the measurements with Power as the y-axis and Bulb Wattage as the x-axis.
52) For the line graph, draw a best fit line and compute slope.
53)
54)   Activity IV : Solar Cell Power due to Wavelength of Light
55) In this activity use one bulb type (60 W recommended) and placed in the lamp at a constant distance (25 cm to 50 cm recommended).
56) Begin with the measurements of the solar cell for voltage and current with no filter in place.
57) Between each trial turn off the light and set up for the next set of measurements.
58) For each of the trials, place one of the color filters (red, green, blue) on the photovoltaic cell and then activate the light and take measurements of voltage and current with that filter in place.
59) For each of the trials, white light and the filter, calculate the Power being produced by that light as measured.
60) Compare results to each other (one way is to let the white light be the 100% base and divide this value into each of the others to find its percentage).
61)
62)   Activity V : Solar Cell Power when connected in Series
63) In this Activity, 3 Solar Cells are connected to each other in a Series Circuit fashion (that is there is one path for the current to flow). To do this : use the alligator clip wires and connect the positive terminal of one solar cell to the negative of the next in line, then its positive to the negative of the next.
64) In the final step have one alligator clip attached to the first solar cell's negative terminal while another alligator clip is attached to the last solar cell's positive terminal. These are the terminals to measure from with the multimeter.
65) Choose a bulb wattage to begin with at a constant distance (25 cm to 50 cm recommended).
66) Choose a way to configure the solar cells so that they receive approximately the same amount of light (examples : 1) they can lie flat in a circle around the light if the wires are long enough, 2) the best choice is to stand them up as we did in Activity 3 since we are going to not only compute this Power but also compare this outcome to Activity 7 when connected in Parallel).
67) Once set up, activate the light and take measurements of voltage and current using the multimeter. For the measurements record the overall circuit and each of the solar cells in turn as well.
68) Note : Be sure to connect the multimeter correctly when measuring voltage and current. For example – the voltage can be done by touching across each set of terminals in the array, while the current is found by connecting the ammeter setting of the multimeter in series with the circuit at one spot (all of the currents will be the same – try a couple different spots to be sure).
69) Also for comparison purposes, after completing the first data set, you redo this exercise with a different size bulb.

70) Before calculation, what does the data show you? (If properly done, when in Series, the voltage should be the sum of all the voltages).

71) From the data set, calculate the Power of the solar cells connected in series. This will be compared to the Parallel circuit case next.

72)

73)   Activity VI : Solar Cell Power when connected in Parallel

74) In this Activity, 3 Solar Cells are connected to each other in a Parallel Circuit fashion (that is there is a unique path for each of the solar cells current to flow). To do this : use the alligator clip wires and connect the positive terminal of one solar cell to the positive of the next in line, then connect an alligator clip wire between the negative terminal of the first solar cell to the second. Connect the third to the second in exactly the same fashion. Hence all of the positives have a line and all of the negatives have a line, much like a ladder.

75) In the final step have one alligator clip attached to the first solar cell's negative terminal while another alligator clip is attached to the last solar cell's positive terminal. These are the terminals to measure from with the multimeter.

76) Choose a bulb wattage to begin with at a constant distance (25 cm to 50 cm recommended).

77) Choose a way to configure the solar cells so that they receive approximately the same amount of light (examples : 1) they can lie flat in a circle around the light if the wires are long enough, 2) the best choice is to stand them up as we did in Activity 3 since we are going to not only compute this Power but also compare this outcome to Activity 6 when connected in Series).

78) Once set up, activate the light and take measurements of voltage and current using the multimeter. For the measurements record the overall circuit and each of the solar cells in turn as well.

79) Note : Be sure to connect the multimeter correctly when measuring voltage and current. For example – the voltage can be done by touching across each set of terminals of any member in the array, while the current is found by connecting the ammeter setting of the multimeter in series with the circuit at one spot – that is be a part of the path for a given solar cell (all of the voltages will be the same – and the currents should be the same too since they are the same solar cell, but this is not always the case - try a couple different spots to be sure).

80) Also for comparison purposes, after completing the first data set, you redo this exercise with a different size bulb.

81) Before calculation, what does the data show you? (If properly done, when in Parallel, the circuits overall current should be the sum of all the currents of the separate solar cells).

82) From the data set, calculate the Power of the solar cells connected in series. This will be compared to the Series circuit case in the last Activity.

83) Notes for all : When constructing Bar Graphs, note that the area of the bar when the axes are Voltage and Current will become Power!

Data :

Activity I : Solar Cell Power due to Exposed Area of Cell

Bulb Wattage Used : _____ W
      Recommended : 60 W

Area of Photovoltaic Cell : _____ cm^2

Distance of Light : _____ cm
      Recommended : 25 cm to 50 cm

% Exposed	Area Exposed (cm^2)	Voltage (V)	Current (A)
100			
75			
50			
25			
0			

Activity II : Solar Cell Power due to Angle of Light Exposure

Bulb Wattage Used : _____ W
      Recommended : 60 W

Distance of solar cell nearest base to lamp : _____ cm
      Recommended : 25 cm to 50 cm

Angle ($^\circ$)	Angle to Light ($^\circ$)	Voltage (V)	Current (A)
0 (horiz.)			
15			
30			
45			
60			
90 (vertical)			

Activity III : Solar Cell Power due to Light Intensity

Distance of Light Source : _____ cm
  Recommended : 25 cm to 50 cm

Wattage Used (W)	Voltage (V)	Current (A)
25		
40		
60		
75		
100		

Activity IV : Solar Cell Power due to Wavelength of Light

Bulb Wattage Used : _____ W
  Recommended : 60 W

Distance of Light Source : _____ cm
  Recommended : 25 cm to 50 cm

Wavelength color	Voltage (V)	Current (A)
White		
Red		
Green		
Blue		

Activity V : Solar Cell Power when connected in Series

Distance of Light Source : _____ cm
  Recommended : 25 cm to 50 cm

Wattage Used (W)	Voltage (V)	Current (A)
25		
40		
60		
75		
100		

Bulb Wattage Used : _____ W

Solar Cell	Voltage (V)	Current (A)
1		
2		
3		

Activity VI : Solar Cell Power when connected in Parallel

Distance of Light Source : _____ cm
          Recommended : 25 cm to 50 cm

Note : This table is for the overall circuit values and not for the individual solar cells in the array. The table below that is for each of the solar cells.

Wattage Used (W)	Voltage (V)	Current (A)
25		
40		
60		
75		
100		

Bulb Wattage Used : _____ W

Solar Cell	Voltage (V)	Current (A)
1		
2		
3		

Calculations :

Se sure to use your Slide Rule !

Area of Solar Cell : **A = L * W**
(rectangular area : Area = Length x Width)

Slope : $m = \dfrac{\Delta y}{\Delta x}$

90° = Angle(1)+ Angle(2)

Electrical Power : (P is in Watts)

P = V*I

Conclusion :

The conclusion depends on which of the Activity(ies) were undertaken. In all cases, there is a calculation of Power and it is examined as compared to some other changing variable (light intensity, distance, et al). What do the results show and how do they compare to science-based expectations?

Summary and Alternate Ideas :

Besides using a constant light source as we did here, try and use the Sun. Be sure to understand its changing position throughout the day and other factors that may affect outcome as well. This makes for a good comparison to the light bulbs used as well.

Activity #27
Make a Radio and Measure and its Signal Activity
Grade Level : High School
Math Level : Calculating

What we refer to as Radio is shorthand for radiotelegraphy from the times when it was called 'wireless telegraphy' and called 'wireless' in Britain. The shortened name 'radio' came from the late 1800s in France from French physicist Edouard Branly and the word radioconductor and picked up by the United States Navy in 1912 and was used often in the military. The first commercial broadcasts began in the 1920s in the U.S. (the first AM radio news broadcast was on August 31, 1920 in Detroit, Michigan at the station 8MK which is still operating today as an all-news format station WWJ and owned by CBS network).

The history of the radio has many inventors, scientists, and engineers the world over and is rather complex. Though the chief contenders for fame claim are Tesla and Edison, there were many physicists from England and Russia as well. (For example, in 1878 David E. Hughes recognized that sparks could be heard in a receiver when he was experimenting with a carbon microphone which was demonstrated to the Royal Society as early as 1880). Initially Marconi had the patent awarded to him. Eventually, however, in a 1943 Supreme Court ruling the term radio was defined by this decision ("A radio communication system requires two tuned circuits each at the transmitter and the receiver, all four tuned to the same frequency") and awarded the first patent to Nikolas Tesla (from 1897) though it was after Thomas Edison's application for a patent in 1885 which was purchased by the Marconi Company in 1891. Tesla had created the foundation devices as early as 1893 and presented them publically at the Franklin Institute in Philadelphia while addressing the National Electric Light Association. While Marconi had demonstrated the basic format of the radio in 1895 and transmitted a signal over 1 mile. From this it was found that the transmission range is proportional to the square of the antenna height and is called "Marconi's Law".

Radio is the transmitting of electromagnetic signals the modulation (changing) of these electromagnetic waves through free space whose frequencies below that of visible light. Like all electromagnetic waves, radio waves travel at the speed of light (3×10^8 m/s). Like all electromagnetic waves this form of radiation travels by the process of oscillating electric and magnetic fields at right angles to each other and at right angles to their direction of travel. In radio waves, the information is carried by the process of systematically changing (the term used is modulating) some property of these waves, such as amplitude, frequency, phase, or pulse width. When these waves come in contact with an electrical conductor (such as in the antenna) the electrically oscillating fields induce an alternating current in the conductor. This can be detected and then transformed into sound or other signals that carry information (i.e. the radio).

All radio systems have these elements : Transmitter, Receiver

A Transmitter, that has a source of electrical energy and this source generates alternating current of an intended frequency of oscillation. The transmitter has a system designed to change (modulate) some property as noted above (amplitude, frequency, phase, etc) of the energy produced to impress a signal on it. The transmitter sends this modulated electrical energy signal into an tuned resonant antenna which converts the changing electrical current into an electromagnetic wave and emits it to free space.

The emitted electromagnetic waves can travel directly or have their path altered through the processes of reflection, refraction, or diffraction. The intensity of the waves falls off as the inverse-square of the distance from the transmission source (see Inverse-square Law of Light Activity) as well as some of the energy being absorbed by various materials in the environment.

The waves are then intercepted by a tuned receiving antenna. This antenna will capture come of the energy of the wave and changes it back into an oscillating electrical current. The Receiver then demodulates these currents. This is the conversion of the electrical current signals into a usable signal form by a detector sub-system. The receiver is tuned so that it responds preferentially to the desired signals (hence not locking onto other signals) (i.e. selecting the station on the radio).

The early radios relied only on the energy collected by the antenna to create signals for the operator. Later invented devices, such as the vacuum tube, and the later transistor makes it able to amplify these weak signals.

Radio systems are not merely portable players and devices in our cars for AM and FM radio stations, but also includes walkie-talkies, radio-controlled remote controlled devices, used in the control of space probes and vehicles, communication in space, navigation, RADAR (Radio Detection and Ranging), cellular phones, along with other broadcasting such as TV which uses AM and FM signals.

What is AM and FM radio?

Both AM and FM are radio signals. AM stands for Amplitude Modulation where the amplitude of the transmitted radio signal is made to be proportional to the sound amplitude that is captured by the microphone, while the transmitting frequency remains unchanged. AM radio signals are affected by static and interference. Natural phenomena, like lightning, have radio emissions at the same frequency as these signals. AM radio stations broadcast with power levels as high as 500 kW originally (today in the U.S. and Canada are limited to 50 kW) and due to reflection off the ionized part of the atmosphere can be literally picked up world-wide.

FM radio signals, stands for Frequency Modulation, where there is amplitude variation at the microphone which causes the transmitter frequency to fluctuate. FM signals do not have the static and interference that often occur with AM signals since it is in a higher frequency area (VHF 30 MHz to 300 MHz) and have transmission ranges of 50 to 100 miles. Also they can penetrate buildings more readily than AM signals. At present, government, police, and fire services use narrowband FM frequencies.

The general idea of the simple Crystal AM radio :

Crystal Radio sets are AM radios and have been around for over half of a century and are essentially the essence of a radio in its simplest form. It has no power source, such as batteries, and only requires a few parts. The power source turns out to be the radio waves themselves that create electrical potential differences   (voltage) and since associated with conductors (wires composed of electrically conducting materials) current can flow.

The basic parts are : Earphone, Ground Wire, Antenna Wire, Inductor (coil of wire), Germanium Diode, and a Variable Capacitor (here in our Activity, Aluminum Foil – the variable capacitor is used to tune in a given frequency (i.e. the station)).

The Antenna is the material (here a wire not grounded) is a conductor that receives the radio waves and converts them into an electrical current. The electrical current flows into the radio circuit between the coil of wire (inductor) and the variable capacitor (aluminum foil). The inductor changes the speed of the electrons by slowing them down. The inductor is a coil of wire, the more coils the longer the path for the electrons, hence the greater the amount of time for the current to change. At the same time it is in parallel to the variable capacitor has a given value of electrical capacity. This creates a back and forth exchange between the coil and the capacitor so that the signal resonates at a given frequency. This frequency is associated with the radio wave frequency for the given station. This signal now flows through the diode which allows only the peaks in the current to flow through and allows it to flow only in one direction to the earphone (whose other wire is connected to the ground wire to complete the circuit) where there is an electromagnet (see the Electromagnet Activity) so that the current flows in a changing manner through a coil of wire wrapped around a magnet. The current affects the strength of the electromagnet which will attract the diaphragm of the earpiece (which is imbedded with iron   particles) so that it vibrates to produce sound.

The Activity is to construct just such a radio and measure the Power of the signal received with a multimeter (voltage and current readings).

Note : Do not do this Activity without parental permission and supervision. Do not use wires, electrical devices, and the like in an unsafe manner. This Activity does not involve the use of any electrical outlets, power sources, and the like. One of the critical aspects of this project is that one must have reasonably close and strong AM radio signals for it to be effective. Also the weaker the signal, the longer the antenna needs to be. There are other parts that can be used and with some research, one can find on the internet other configurations and materials to use for the crystal radio set. For example, you can create a variable inductor instead of a variable capacitor. There are even kits already prepared for construction and use.

Purpose : To construct a basic AM Radio from common materials, receive local radio station signals, and measure the power of the signal received through readings of voltage and current on a multimeter.

Materials :

- Aluminum Foil,
- Magnet Wire (22 or 26 gauge & length),
- Wire (20 gauge, 10 m long, will be cut – antenna & ground),
- Wires with alligator clip ends (to make assembly easy),
- Poster Board (can act as base for radio),
- Wire Stripper (if needed),
- Sandpaper,
- Multimeter,
- Earphone,
- Diode (1N34A) (Radio Shack pt no. 276-1123),
- Large Plastic Cups that can nest in each other (20, 24 or 32 oz),
- Masking Tape and Clear Tape,
- Note : Need to live an area where AM radio signals are available,
- Slide Rule

Note : The larger the cups used, the better. They can also be Styrofoam or Paper. Also the construction may involve the creation of several of these (so as to have different capacitors).

Note : Measuring Voltage and Current requires parental permission and supervision. Also these electrical measurements are only done with this type of radio and no other form! As with all electrical activities, exercise caution in using the tools properly and safely.

Set Up :

- The longer the Antenna wire, the better. One could use a wire that they connect to an antenna (car antenna, or an old TV antenna, hence the wire need not be as long. If the wire is the antenna, then it needs to have some length and be separated from the ground).
- Use the magnet wire and wrap it at least 100 times (or more as desired) around the paper towel tube and leave long enough ends so that it can be attached to the circuit to make the radio.
- Use sandpaper to remove the coating from the magnet wire ends.
- Cover half of the cylinder nearly completely around of one of the plastic cups. Tape with clear tape the foil in place. Make sure it is smooth and flat to the surface.
- Cover the other cup cylinder surface with aluminum foil as well ½ way around so that the same amount of area is covered.

- Use Clear Tape and tape to each of the cups a wire where the ends have been stripped. One wire per cup. Nest one cup in the other. The bare ends of the wire are out of the cup.
- Use alligator clip wires and attach one of the magnet wires to one of the two cup wires to the clip of the wire in use.
- Connect a second alligator wire to the other wire from the other cup and the other end of the magnet wire.
- The above description means that the Coil (the magnet wire tube) and the Variable Capacitor (the Aluminum Foil covered Cups) are in parallel to each other (see photos of items and diagrams below).
- To one of the free alligator clip wire ends attach the diode. (We will call this end 'A' since one more attachment will be made to it!)
- To the diode, attach one of the wires of the earphone.
- Attach the other earphone wire to the other free alligator clip wire end. (We will call this end 'G' since one more attachment will be made to it!)
- To end 'A' attach the wire (or item) that is to become the Antenna.
- To the end 'G' attach a wire to become the ground, which can be a cold water faucet. Note : In either case of the Antenna or Ground they are not attached to electrical outlets or devices and you have supervision for such actions.
- The Radio should be operational now, but realize you may have to adjust the cups (your variable capacitor) and move the antenna plus check your connections. With patience and effort, you should be able to obtain a radio signal (provided you live within range of a strong enough signal).
- Now move on to the measurements in the Procedure.

Photos from Set Up :

 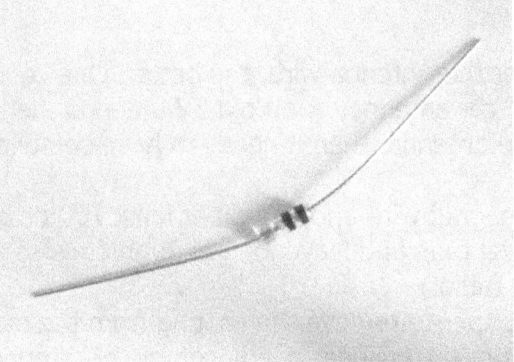

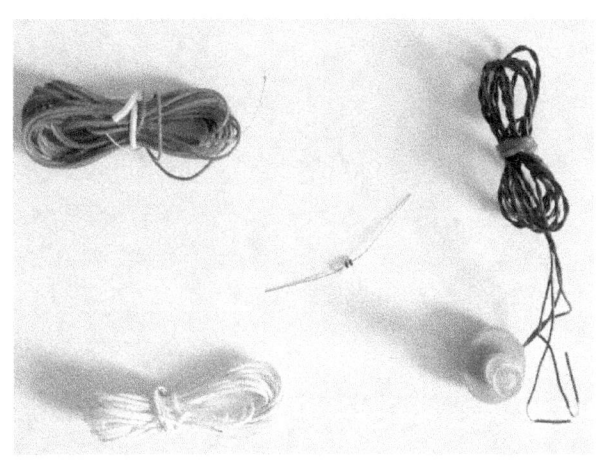 

261

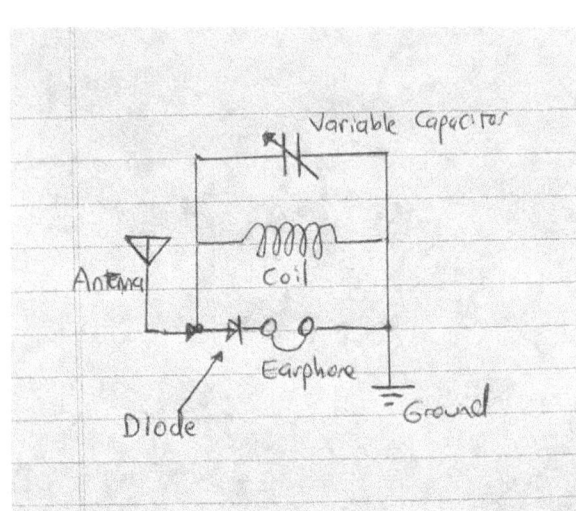

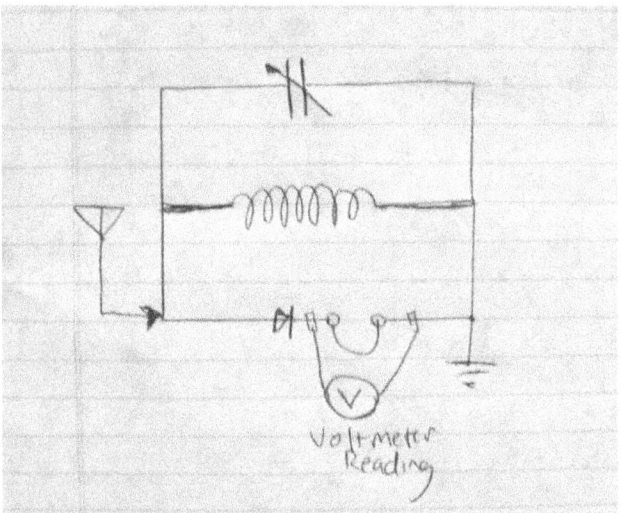

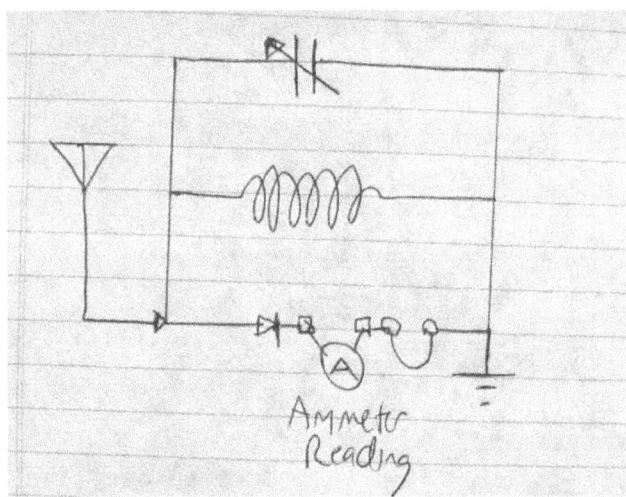

Procedure :

- The procedure follows the construction and testing of the Radio as described in the Set Up.
- With the Radio operational do the following :
- Listening with your earphone find 2 or 3 radio stations and write down their call letters and frequency in the data table.
- For each radio station do these steps to measure the amount of voltage and current in your system as caused by the radio station signal through your radio :
- For both the Voltage and Current readings you can use a digital or a needle-based multimeter. In either case, it is set in DC settings.
- Note that the overall readings are strongly connected to the distance of the radio station signal source and its strength. The farther the source, the smaller the readings.
- Note that for Voltage and Current there are different set up configurations needed to measure the power of the radio signal.

- In both the Voltage and Current cases, the values are very small.
- The DC Voltage setting should start at regular voltage and then dial it down with each step to the millivolt setting.
- For Voltage it is set up across the earphone wires so that it is in parallel with the system.
- Typical values are from 15 mV to 225 mV (but depends on distance).
- Disconnect the multimeter and change the settings to DC Current reading.
- You can start with Amps, but it will quickly go to milliamps (mA) and then microamps (μA) readings.
- Next disconnect the ground wire connection of the earphone and connect it in series to the circuit the multimeter and then to the ground so that the multimeter is now part of the continuous circuit.
- As noted above, start with Amp readings, then move to milliamp readings and then to microamps readings.
- The longer the antenna will also affect the reading values, but they can be from 2 μA and up to 200 μA.
- Record both the Voltage and Current readings for each of the Radio stations discovered on your radio.
- For calculations (being wary of the proper decimal placement) calculate the Power of the Radio Signal in each case.
- Though the Slide Rule is a recommended tool, all of these calculations can be done with a regular or scientific calculator. Some scientific ones even have built-in averaging formulae. For those who like spreadsheets, the data can be typed in and the formulae then also be typed in its own cell where the formula references each of the measured variables in their respective cells, for example B1..BN has the measurements and values used in the equation while BN+1 has the formula for all of these variables (why not A? Simple – use it to label you variables)
- For further investigation, find out how far away the stations are and the transmission power of the signal to see how this compares to your calculated power values (that is to say are similar signal transmission emissions at different distances different in power – the farther the smaller the power value, for example?)
- Other things to try : Change the number of coils in your Coil or the size of your Capacitor. Change the length of the Antenna. Try other materials as well.

Data :

Length of Antenna Wire (L) : _____ (m)
Number of Coils on Inductor (N) : _____

Trial	Radio Station (λ or station)	Voltage (mV)	Current (μI)
1			
2			
3			

Calculations :

Be sure to use your Slide Rule!

Measured Power for a Station received by the Radio :

P = V*I

Conversions Needed :

μ means 10^{-6}
m means 10^{-3}
1 m = 100 cm

$1 V = 1 \times 10^3$ mV
$1 A = 1 \times 10^6$ μA

Conclusion :

What values did you measure and compute for the power of stations received by your radio? Were they significantly different from each other? If so, why do you think this is the case? (Look up the radio stations and find out how much power they use in broadcasting their signal and how far away from you they are). (Note : Most are 50 kW).

With regards to the radio – you can try different size variable capacitor systems, a different number of coils on your inductor and/or a different length of an antenna. Do any of these affect the performance of the radio and the amount of power measured from your readings for a given station?

Notes : Power readings will be very small and can be from 1×10^{-5} to 1×10^{-8} W (I have personally done this with the items noted above in the photos and measured 2.7 mA and 67 mV for a power reading of 1.8×10^{-7} W)

Activity #28
Rate of Radiometer Rotation Activity
Grade Level : High School
Math Level : Calculating

A Radiometer is any device that responds to electromagnetic radiation through a physical motion. The most common is called a Crookes Radiometer which we see today as a 4-vane system where each vane has either white (silver) paint or black paint on the opposite side. The system rotates atop a point in a bulb that is slightly evacuated. The bulb is clear. Light entering the bulb strikes the vanes and the device will rotate in response to the intensity of electromagnetic radiation striking it. The greater the intensity of light, the faster the rotation. Hence it could be used to measure the intensity of radiation in an indirect way. The only problem is to measure the rate of rotation of the Radiometer. To do this it is best to use a digital tachometer.

The Crookes Radiometer is one of the earliest versions of radiometers, made by Sir William Crookes in 1873 during his investigations of infrared radiation. Today's versions are comparable and called a light-mill or solar engine. Light the early versions, the rotation makes it seem that light striking the hotter dark side is pushed and so rotates towards the lighter sided vanes.

Many ideas have been proposed over its operation, even by persons such as James Clerk Maxwell, the father of electromagnetism. His explanation was radiation pressure, but this fails to explain the idea and in fact would predict an opposite rotation since the black side should absorb the energy while the white or silvered sides should reflect the energy hence there should be a net pressure twice as great on the white sides to cause it to spin in the opposite direction.

None-the-less, the black sides do in fact absorb more energy. Now there is the need for an explanation for the rotation. The correct answer came from Osborne Reynolds in 1879. In a paper submission to the Royal Society he described what he called 'thermal transpiration'. In it he meant the flow of a gas through porous plates caused by temperature differences on the two sides of the plates. On opposite sides of a vane the pressure is initially the same, but in time there is a temperature difference or gradient, where now the gas will flow from the colder side to the hotter side, resulting in a net pressure on the hotter side. Equilibrium is reached when the ratio of pressures on either side is the square root of absolute temperatures. This counterintuitive result is due to tangential forces between the gas molecules and the sides of the narrow pores in the plates.

Here the vanes are not porous, however. To find the solution, one needs to consider the edges of the vanes. The effect is then called 'thermal creep'. Here the gases creep along the surface that has a temperature gradient. These, though, are the same thermomolecular forces as described by Reynolds in his thermal transpiration. The net movement of the vane due to the tangential forces around the edges is away from the warmer gas and towards the cooler gas, with the gas passing around the edge in the opposite direction! Essentially the cooler gas is coming around the corner from the white side to the warmer gas on the black side. The behavior is just as if there were a greater force on the black vane side of the given vane.

Despite this interesting conclusion, the radiometer is a simple and fun thought item and can be used to find such considerations as : how does a given light intensity at varying distances affect the rate of rotation of the radiometer? How does the difference in light intensity at a constant distance affect the rate of rotation of the radiometer? These are explored by using a light source, a measuring tape, a digital tachometer, and a radiometer, of course. We will calculate average readings, and the slope of the graphs of average rotation rates versus things such as distance or light intensity to see what sort of linear or exponential relations exists in these cases. (To see why a exponential relation might exist look at the Inverse-Square Law of Light Activity).

Purpose 1 : To determine the relationship between the distance of a light source from a radiometer and its rate of rotation

Purpose 2 : To determine the relationship between the intensity of light at a given distance from a radiometer and its rate of rotation

Purpose 3 : To determine the relationship between the type of light at a given intensity and distance from a radiometer and its rate of rotation

Materials :

- Radiometer,
- Digital Tachometer (non-laser type),
- Meter Stick or Measuring Tape,
- Small Desk Lamp,
- Several Bulbs (a group of 3 of the following): 100 W, 75 W, 60 W, 40 W,
- Other Types of Bulbs for comparison : Incandescent bulbs with the same wattage (if using halogens in the prior list or vice versa),
- Timer – can be a watch, clock, or stopwatch,
- Slide Rule

Safety Notes : In the case of all bulbs used : do not touch, place objects on or too near the bulbs, and do not look into any of the lights when active. Be sure to have parental permission and supervision when handling them and doing any of these Activities. When exchanging bulbs, exercise caution – allow a lot of time for cooling – for example.

Procedure :

1) All of the Activities use the same basic tools : Light Source, Radiometer, Digital Tachometer, Measuring Tape, Timer, and Slide Rule. The only difference will be the following : 1) varying Distances of a constant Light Source, 2) varying Intensity of similar Light Sources at a constant distance, and 3) Different Types of similar Intensity Light Sources at a constant distance

2) For all Activities be sure to have the Light Source under consideration be the only source of power for the Radiometer – this might mean pulling the shades, having other lights off, et al.

3) Note that if using a halogen bulb, like fluorescent bulbs, the digital tachometer will respond to it (in 2 blade mode it will see it as a 4800 rpm object), so it is best to use incandescent for the activities. If no other choice - to compensate for this problem be sure to point the tachometer away from the light source as best as possible and have a barrier (book, other item) to block the light from the tachometer.

4) For all Activities the Digital Tachometer should be at the same distance from the Radiometer at all times. It is best to test its operation beforehand.

5) It is best to have the Digital Tachometer rest atop a stack of books or a inverted plastic crate tall enough so that it is in line with the blades of the Radiometer. Choose a distance for it to be at for all Activities and all Trials in the Activities. (suggestion depends on the range of the device, but it is typically 3 cm to 9 cm (from the bulb)).

6) Important Note for All Activities : The Digital Tachometer has normally 2 settings – 2 blade and 3 blade – if you can find a 4 blade one then use it – if yours is 2 or 3 blade, then set it on 2 blade. But realize that the radiometer has twice as many blades (4) – so you need to adjust your measurements accordingly. You need to divide its reading by 2.

7) In bulb selection be sure to test the Radiometer with that bulb on in at least 2 situations – close and far – to see if it is operating effectively. If not, be sure to adjust your procedure distances and bulb intensities accordingly.

8) At no time do you stare into the bulb. Do not touch the bulb when on or even after it is off – they can be very hot. If you have to change the bulb (such as Activity 2 and Activity 3) then have a parent involved to do this – be sure to let it cool for some time after use.

9)

10) <u>Activity 1 : Distance of Light Source vs. Radiometer Rotation Rate :</u>

11)

12) In the first Activity, it is best to choose a sufficient intensity bulb, for example 75 W or 100 W as this will provide the best response and can affect the operation of the radiometer at the greatest range of distances chosen. (Be sure to note the maximum power for the lamp chosen and do not exceed this ! Most are either 40 W or 60 W – try to find one that can use a 75 W if possible).

13) Next select the distances to be used – which is best to choose ones that are multiples of each other (can be even or not). An example might be 8 cm, 16 cm, 24 cm, 32 cm (all measured from the bulb). – This selection depends on the bulb and the response of the radiometer.

14) With distances [d] chosen, place these in the Data Table.

15) With good choices of distances, you can choose to then generate a Relative Distance [r_d] list. This is done by using the smallest distance as the divisor and dividing it into all of the values (including itself), so that the Relative Distance will result in a table of whole number values, such as 1, 2, 3, et al.

16) Note that Relative Distance is just a convenient way to examine results and not an absolute necessity in the Activity. So you have to decide – use your Distance values as they are, or use the Relative Distances when examining the data – both produce the same outcome.

17)

18) <u>The Activity 1 : Distance of Light Source vs. Radiometer Rotation Rate</u>

19)

20) We will start with the assumption (as noted in Activities like the Inverse-Square Law of Light) that the Radiometer should receive the least amount of power with the Light Source at its greatest Distance, so start there.

21) With each trial, have the Radiometer set up, the Digital Tachometer activated, and the Light Source on. Give the system about 2 minutes to allow the Radiometer to begin rotating, then take a reading from the Digital Tachometer.

22) Decide on the amount of time for the subsequent readings – say each 1-2 minutes apart – and take these for this given Trial.

23) Be sure to use the same amount of time for these readings so as to maintain consistency from one Trial to the next. This is why you have a timer of some sort.

24) For each reading, always let the system sit and stabilize for a few seconds. The values may oscillate around a couple of numbers – take the one that seems to be the most frequent.

25) Between each of the Trials, turn off the light, let the Radiometer come to rest, and then move the Light Source to its new Distance. Then redo the Trial measurements.

26)

27) Calculations :

28)

29) First, determine the Average Rotation Rates for each of the Trials.

30) Next graph the Rotation Rate [R] on the Y-axis and either the Distance [d] OR the relative distance [r_d] on the X-axis. Draw a curve through this. Hint : Note it might not be a straight line.

31) If approximating this line as linear, you can determine its slope if you want. If you think it may not be linear, then employ the next steps so as to determine the exponential relation between Rotation Rate and Distance.

32) Next Generate a Table of the log values of the aforementioned points. Here we want log(R) and the log value of 'd' or 'rd'. (log(d) or log(rd)) using the Slide Rule and the D Scale with the L Scale.

33) Now graph these points log(R) on the y-axis and the log(d) on the x-axis. Draw a best fit line and determine the slope of this line. Hint : It should be linear. Hint : the slope should be -2 ideally which indicates an inverse-square relation between the variables.

34) A further check of the Inverse-Square Relation can be found by doing the following :

35) First use the D Scale looking up either the distance (d) or the relative distances (rd) used in the Activity and find the corresponding value on the C1 Scale (its inverse). Be sure to keep track of the decimals here! Jot these down someplace.

36) Look up the square of these inverse values by looking them up on the D scale and finding its square on the A scale. Be sure to watch your decimal values. Use this information and fill in this portion of the Data Table.

37) Graph these data points where the Intensity per Unit Area values (R) on the y-axis vs. the inverse-square relative distance values you have calculated ($\frac{1}{d^2}$) on the x-axis. Draw a straight line through these data points and determine the slope of this line (which should be 1).

38)

39) Extended Activity for Activity 1 : Other bulbs for comparison –

40)

41) If you choose to conduct further experiments for other bulbs other than the one chosen in this Activity, create separate tables for each of the bulbs chosen. In the matter of your Data Table distances (d) it is best to set up a chart of the same chosen distances used.

42)

43) Activity 2 : varying Light Intensity vs. Radiometer Rotation Rate :

44)

45) In this Activity you are using the same materials noted above in the procedure reading.

46) You need to select a constant distance and record this on your Data Table. All bulbs used will be at this distance for each of the Trials.

47) It is then best to determine which selection of bulbs to use. Minimally 3 is best (such as 40 W, 60 W, and 100 W for example).

48) Note that the range of bulbs used does depend on the lamp being used and its rating. If it can only handle a 75 W bulb at most then that is the upper limit used. Also as to the lower limit, if you choose a distance too far away from the bulb, then a low intensity bulb, such as a 40 W one, might not promote a response from the radiometer, so be sure to test the chosen distance with the least powerful bulb to see if it can cause the radiometer to spin.

49) It is best to use the same type of bulb (such as only halogen or only incandescent) since Activity 3 explores the question of different types of bulbs. Here we are examining the effects of a given Light Intensity at a constant distance and its effect on the Rate of Rotation of the Radiometer.

50) In the Trials it is best to start with the lowest wattage and then move on to the higher ones in sequence.

51) For each Trial – activate the light at the given distance and allow the radiometer to begin working.

52) At a designated time (say 1 or 2 minutes) you take a Tachometer Reading and record it.

53) Have 3 readings per Trial and have the same amount of time between readings (such as 2 minutes).

54) The 3 readings for a given Trial are to be Averaged into a single value to be used in the Graphing of the Data.

55) Between each of the Trials it is important to note the following :

56) First, since the bulb has been on during the Trial, it is hot – so do not remove it right away and if needed and deemed necessary, let your parents remove the bulb when it is time. It needs to cool sufficiently. Always act in a safe manner with items like this.

57) Also be sure to have the lamp turned off – as an added safety precaution, you should unplug it – when exchanging bulbs between Trials.

58)

59) Calculations :

60)

61) First, determine the Average Rotation Rates for each of the Trials.

62) Next graph the Rotation Rate [R] on the Y-axis and Wattage of the bulbs from least to greatest on the X-axis. Draw a curve through this. Hint : Note it might not be a straight line.

63) If approximating this line as linear, you can determine its slope if you want. If you think it may not be linear, then employ the next steps so as to determine the exponential relation between Rotation Rate and Light Intensity (Wattage).

64) Next Generate a Table of the log values of the aforementioned points. Here we want log(R) and the log(W) using the Slide Rule and the D Scale with the L Scale.

65) Now graph these points log(R) on the y-axis and the log(W) on the x-axis. Draw a best fit line and determine the slope of this line. Hint : If there is a relationship that is exponential, the line should be linear. The slope will reveal the exponential relationship.

66) Of the relationship is an Inverse-Square one, use the Calculations section from Activity 1 for further analysis of the data. If not, then your data analysis is completed.

67)

68) Activity 3 : different Light Sources vs. Radiometer Rotation Rate :

69)

70) In this Activity you are using the same materials noted above in the procedure reading.

71) You need to select a constant distance and record this on your Data Table. All bulbs used will be at this distance for each of the Trials.

72) It is then best to determine which selection of the types of bulbs to use. Minimally 2 is best (Note : It is best if they have the same rating in wattage such as an incandescent bulb and a halogen bulb of 60 W, for example).

73) Note that the wattage of bulbs used does depend on the lamp being used and its rating. If it can only handle a 75 W bulb at most then that is the upper limit used. Also there is a lower limit to a bulb being used, if you choose a distance too far away from the bulb, then a low intensity bulb, such as a 40 W one, might not promote a response from the radiometer, so be sure to test the chosen distance with the least powerful bulb to see if it can cause the radiometer to spin.

74)

75) In the Trials it is best to have the same procedure for each of the bulbs. It does not matter which you start with.

76) For each Trial – activate the light at the given distance and allow the radiometer to begin working.

77) At a designated time (say 1 or 2 minutes) you take a Tachometer Reading and record it.

78) Have 3 readings per Trial and have the same amount of time between readings (such as 2 minutes).

79) The 3 readings for a given Trial are to be Averaged into a single value to be used in the Graphing of the Data.

80) Between each of the Trials it is important to note the following :

81) First, since the bulb has been on during the Trial, it is hot – so do not remove it right away and if needed and deemed necessary, let your parents remove the bulb when it is time. It needs to cool sufficiently. Always act in a safe manner with items like this.

82) Also be sure to have the lamp turned off – as an added safety precaution, you should unplug it – when exchanging bulbs between Trials.

83)

84) Calculations :

85) First, determine the Average Rotation Rates for each of the Trials.

86) Next graph the Rotation Rate [R] on the Y-axis and Type of Bulb on the X-axis. Note that this is a Bar Graph.

87) The only calculations done in this Activity are the Average Rate values.

Data :
Activity 1 : Distance of Light Source vs. Radiometer Rotation Rate

Wattage of Bulb Chosen for Activity : _____ W

Trial	Distance [d] (m)	Relative Distance [r_d]	RPM rate reading 1	RPM rate reading 2	RPM rate reading 3	Average RPM rate

Note : Choose to use the Distance (d) or Relative Distance (r_d) and fill in one of the following tables :

Distance (cm)	$\frac{1}{d^2}$	Average RPM rate [R]

Tabulated Data Table to Examine Inverse-Square Relation Validation :

Relative Distance [r_d]	Inverse-square of Relative Distance ($\frac{1}{(r_d)^2}$)	Radiometer Rotation Rate [R]

Table of Log-Log Values to Graph
> Note : Choose to have created a Relative Distance value or not. Either can
> be used (relative distance or distance) – choose only 1.

Log (Relative Distance) or Log (Distance)	Log (Radiometer Rotation Rate)
Log [r_d] or Log [d]	Log [R]

Activity 2 : Light Intensity (Wattage) vs. Rotation Rate

Constant Distance for All Light Sources : _____ m

Trial	Light Intensity [W] (W)	RPM rate reading 1	RPM rate reading 2	RPM rate reading 3	Average RPM rate [R]

Table of Log-Log Values to Graph – Test to see if the relation is linear or not

Log (Wattage)	Log (Radiometer Rotation Rate)
Log [W]	Log [R]

Activity 3 : Type of Light vs. Rotation Rate

Constant Distance for All Light Sources : _____ m

Trial	Type of Light [T]	RPM rate reading 1	RPM rate reading 2	RPM rate reading 3	Average RPM rate

Calculations :

Be sure to use your Slide Rule!

Be sure to adjust the Digital Tachometer reading if necessary.

Formulae :

Average Value :

$$\text{Average Value} = \frac{\text{Sum of all Trials}}{\text{Number of Trials}} = \frac{\Sigma(\text{trials})}{\text{number of trials}}$$

Relative Distance (r_d) :

$$r_d = \frac{\text{measured distance-} d_x}{\text{smallest distance } d}$$

Slope :

$$\text{Slope} = m = \frac{\Delta Y}{\Delta X}$$

$$m = \frac{\Delta \log(R)}{\Delta \log(d)} = \frac{\Delta \log(R)}{\Delta \log(r_d)}$$

$$m = \frac{\Delta(R)}{\Delta(\frac{1}{(d)^2})} = \frac{\Delta(R)}{\Delta(\frac{1}{(r_d)^2})}$$

Conclusion :

In each of the Activities, what did you notice when it came to the rate of Radiometer spin and the measured and controlled independent variable you were considering – such as distance or light intensity? For example, as the distance increased, what happened to spin rate? In the case of light intensity, as light intensity increased, what happened to spin rate? Why do you think this was the case? (For ideas and help, read the Inverse-square law of Light Activity and look at the Solar Cells Activity as well). Were your slopes positive or negative? Were your graphs linear? If they were not linear, were you able to approximate the relation of the variables using the log or L scale of the Slide Rule? What sort of relations did you find? What do your results say about the relation of light intensity and the ability to obtain useful work from it?

274

Worksheet #1
Electromagnetic Wavelength Calculations
Grade Level : High School
Math Level : Calculating

Determining the Frequency of Electromagnetic Radiation from its Wavelength Activity

Einstein and others in Physics long ago showed that the speed of light is constant in a vacuum (3×10^8 m/s approximately). This is true for all wavelengths of the Electromagnetic Spectrum and is not confined to the speed of light. The term speed of light is merely a phrase that really applies to the whole of the electromagnetic spectrum and is used due to our daily familiarity with it.

The Electromagnetic Spectrum is considerable. From the shortest wavelengths to its longest wavelengths are these basic categories :

Gamma Rays, X-Rays, Ultraviolet Rays, Visible Light, Infrared Radiation, Microwaves, Radio Waves.

Many of us know the regular acronym for the Visible Light Spectrum : ROYGBIV.
It represents Red, Orange, Yellow, Green, Blue, Indigo, and Violet. Many today due not refer to Indigo as much, but the term does not sound as nice without the 'I' in it. It was Newton who realized that white light from the Sun is composed of the whole of the color spectrum.

Radio Waves, of course, have the widest array to them and includes TV signals, FM & AM radio, our Cellular Phones, and the like.

All of these forms are fully mathematically described by James Clerk Maxwell in 4 formulae showing that they are each combinations of oscillating Electric and Magnetic Fields.

Through the years, there are many manifestations of these forms in nature, which are studied to help us decode and understand the Universe and any and all of its components. Many devices have been created to make use of them.

This Activity is basic – simply take each of the wavelengths given (if a range do both the lower and the upper value in turn) and determine the frequency of the electromagnetic radiation wavelength.

The way to do it is from the basic equation, known as the Wave Equation :

$$c = \lambda * f$$

Electromagnetic Radiation	Wavelength (m) Symbol : λ	Frequency (Hz) Symbol : f
Radio Waves	$> 10^{-1}$	
Microwaves	1×10^{-1} to 1×10^{-3}	
Infrared	1×10^{-3} to 7×10^{-7}	
Visible Light	6×10^{-7}	
Red Light	$625 - 740 \times 10^{-9}$	
Orange Light	$590 - 625 \times 10^{-9}$	
Yellow Light	$565 - 590 \times 10^{-9}$	
Green Light	$520 - 565 \times 10^{-9}$	
Blue Light	$450 - 500 \times 10^{-9}$	
Ultraviolet	4×10^{-7} to 1×10^{-8}	
X-Rays	1×10^{-8} to 1×10^{-11}	
Gamma Rays	$< 10^{-11}$	

Red Light is 700 nm while Blue Light is 400 nm for most considerations

Worksheet #2
Stellar Distances from Magnitude Calculations
Grade Level : Middle School
Math Level : Calculating

Astronomy Slide Rule Activity
Determining Stellar Distances from Magnitude!

- ➢ In everyday life a given object will appear as bright as it does from the viewpoint of an observer. In Astronomy, the term for this is called the **Apparent Magnitude** (denoted by m). It is the brightness of an object as it is in the sky.
- ➢ Also in Astronomy, a quantity called **Absolute Magnitude** (denoted by M) is used to tell the relative value of the brightness of an object if all of the objects are lined up and at the same distance from the observer.
- ➢ Originally the scale started at 0 and had positive values, but did not include the Sun or Moon. Also historically it had only whole number values. Later with measurements using mechanical devices negative values, such as for the Sun and Moon could be determined and values to the nearest 0.1 are accurately determined as well.
- ➢ So the brighter an object is, the smaller its positive absolute magnitude value is or the more negative its absolute magnitude is.
- ➢ The information given for our example :
- ➢ The Sun has an Absolute Magnitude of -26.7 and the Moon has an Absolute Magnitude value of -12.5. About how many times brighter is the Sun than the Moon?
- ➢ The formula :

- ➢ $R \cong 2.51^{[M1-M2]}$

- ➢ Note (2.51 is rounded from 2.511886--- which comes from the 5th root of 100 since every 5 magnitudes results in a brightness difference factor of 100x)

- ➢ Notice the square brackets. These are absolute value signs. We subtract the values and take the absolute value of them.
- ➢ Here : [-26.7 – (-12.5)] = [-14.2] = 14.2

- ➢ $R = 2.51^{14.2}$

- ➢ Let's take the Log of both sides
- ➢ Log(R) = 14.2*Log(2.51)
- ➢ Realize we do this so as to make the job easier to solve, as this is like the earlier problem of raising a number to a non-whole number value.
- ➢ To solve, first look from 2.51 on the D scale to the L scale value and find the value 0.4

- Next, align the Left Index of C scale on 4 for D scale (think of it as 0.4) and
- Read across the C scale to 1.42 and find an answer on the D scale of nearly 5.68 (the 8 is estimated of course) (see photo)
- Notice I did not round to 5.7 or 5.9. First, Small changes such as this will drastically affect our answer.
- Also note I actually calculated the 3rd digit at 8 since 4x2 = 8 and this would be our last digit.
- The 5 in front of the decimal in our answer is the power of 10 we will multiply our answer by to find the result.
- (We did this same technique in the L scale example above)
- We look up the 0.68 on the L scale and read the corresponding value on the D scale of 4.80, which we multiply by 10^5 and this yields an answer of 479,000.
- The scientific calculator yields an answer of 478,628. For the sake of significant figures, this number would be rounded to 479,000 since there are 3 numbers in our measurement.
- If we assume the calculator is supreme as the true value, then our error is 0.00077.
- Better still is the fact that to 3 places, the slide rule is exactly the same as the once-revered scientific calculator.
-
- The next major question is, though, is there a relation between Absolute Magnitude and Apparent Magnitude and the Distance an Object really is?
- The answer is yes!
- The formula is :
-

$$m - M = 5*\log_{10}\left(\frac{d}{10}\right)$$

-
- Where 'm' is the apparent magnitude, 'M' is the absolute magnitude, and 'd' is the distance from the observer to the object in parsecs.
- Note : A parsec is the distance of an object that has a parallax angle of one second. (Parallax is the apparent shifting of nearby objects with respect to distant ones as the position of the observer changes).
- Here is the formula reworked to solve for distance :
-

$$d = 10^{\frac{m-M}{5}+1}$$

-
- Here is an example :
-
- We will use : the star Rigel found in the constellation Orion.
- Rigel has an apparent magnitude of 0.14 and an absolute magnitude of -7.1
- The difference yields + 7.24 (from 0.14 − (-7.1))
- Using the slide rule and dividing this by 5 we obtain :
- 1.445
- Added to 1 is 2.445
- Now we have 10 to this power.

- ➤ On a standard 9-Scale Slide Rule it is best now to use the L Scale here.
- ➤ The great answer is that the log is the exponent in this case, so
- ➤ The log(d) = 2.445
- ➤ Now look up the mantissa 0.445 on the L Scale and find its antilog on the D Scale :
- ➤ It is 2.78
- ➤ The characteristic of 2 is the power of 10 for this value, so 100*2.78 is 278.
- ➤ What is this answer?
- ➤ **What is 278?**
- ➤ It is the distance in parsecs of Rigel from the Earth.
- ➤ To find the value in the more commonly known light years, the conversion factor is **1 pc = 3.26 ly**
- ➤ Leaving the cursor on the antilog value of the D Scale, move the C Scale Index over it and read along the C Scale to 3.26 and find the answer below :
- ➤ It is 906
- ➤ **This is 906 light years.**
- ➤ Depending on how you round this off, it is about 900 light years, which is in agreement with our table below.
- ➤ *Follow this set of steps and find the distance to the stars merely knowing how bright they are in the sky as we see them and how bright they really are if all placed at the same distance from us.*
- ➤ *Imagine that!*
- ➤ This Activity provides a Table of various stars, some of them well known and very bright, others just nearby. The first table gives the star name, its absolute and apparent magnitudes and the second table is the answer, the distance to these stars, which you can do for yourself, like the example above.
- ➤ Be sure to watch for the signs (+ and -), read the scales carefully and be sure to convert your answers, since the first answer you obtain is in parsecs.
- ➤ 1 parsec = 3.26 light years
- ➤ 1 light year = 5.8×10^{12} mi or 9.5×10^{12} km
- ➤ A light year is the distance that light, traveling at 3×10^8 m/s, would traverse in the course of 1 Earth year.
- ➤ Enjoy.

The Table of Stars and Their Magnitudes

Star	Absolute Magnitude (M)	Apparent Magnitude (m)
Sun	4.83	-26.8
Sirius	1.4	-1.47
Arcturus	-0.3	-0.06
Vega	0.5	0.04
Rigel	-7.1	0.14
Betelgeuse	-5.6	0.41
Spica	-3.3	0.91
Antares	-0.51	0.92
Alpha Centauri	15.5	0.01
Regulus	-0.7	1.36
Deneb	-7.1	1.26
Altair	2.2	0.77
Pollux	1.0	1.16

The Answers in This Table :

Star	Distance (ly)
Sun	1.5×10^{-5}
Sirius	8.7
Arcturus	36
Vega	26.5
Rigel	900
Betelgeuse	520
Spica	220
Antares	520
Alpha Centauri	4.2
Regulus	84
Deneb	1600
Altair	16.5
Pollux	35

Project
Personal Slide Rule Template

MAKE A PAPER 6" SLIDE RULE

On the following two pages are two different templates than enable you to make a 6" slide rule. Make your choice as to which you want to construct. They each have the same number and type of scales and are quite complete and useful. They have these scales for use : C, D, C1, D1, CF, DF, C1F, A, B, S, T, ST, K, L

Things Needed : One of the Templates, Scissors, Ruler

Steps to making a slide rule :

1. In either case make 2 copies of the chosen Template.
2. For both you need something to act as a cursor – best choice is a ruler. Be sure to align it with a straight edge, such as the bottom of the paper in the case of your unfolded slide rule and be sure to fold along a straight line so as to be able to use this for the folded model.
3. In the case of the first slide rule which is the unfolded slide – this is the set of scales where they are all bunched together and in separate boxes do the following :
4. Leave the first copy alone. It acts as the stators for your slide rule and simply lie on the table
5. With the second copy, cut out the slide – the middle set of scales – so that it can be moved along between the stators as needed.
6. It is now ready to use – Use a ruler where it lays across the stators and slide perpendicular to the direction one regularly reads the paper and the bottom edge of the rule is aligned with the bottom edge of the paper (needs to be perpendicular).
7. In the case of the second slide rule which is the folded slide – this the set of scales in sets of 5 in boxes which are separated.
8. As in the first slide rule, make two copies of the template to be used.
9. With the first copy fold it so that the top and bottom set of scales are now opposite the middle set of scales. It is best to fold it so that it would align with the middle set of scales as if it were a slide and the top and bottom are the stators. – Note if there is excess paper above and below the stators so that it would interfere with the slide and their reading, cut this away in a straight a manner as possible (follow the line of the box encasing them).
10. Now fold the second template so that it fits into the sleeve and the middle set of scales faces out of the space between the top and bottom stators and is now the slide.
11. It may take some adjusting, so be patient.

12. Be sure to follow the lines of the boxes for folds as it is critical that the upper and lower stator have aligned scales.
13. With your cursor (best choice is probably a ruler as in the first case) be sure that the end edge aligns with the lines and the ruler itself is the cursor line.

Some notes for Use :
In order to use the A/B or CF/DF scales these are on the adjacent slide and stator in a doubled fashion.
Have fun and enjoy :)
Thanks :
These scales came from the web site : The International Slide Rule Museum (sliderulemuseum.com) found on the Slide Rule Reference Scales tab and set with graphics by Andrew Nikitin

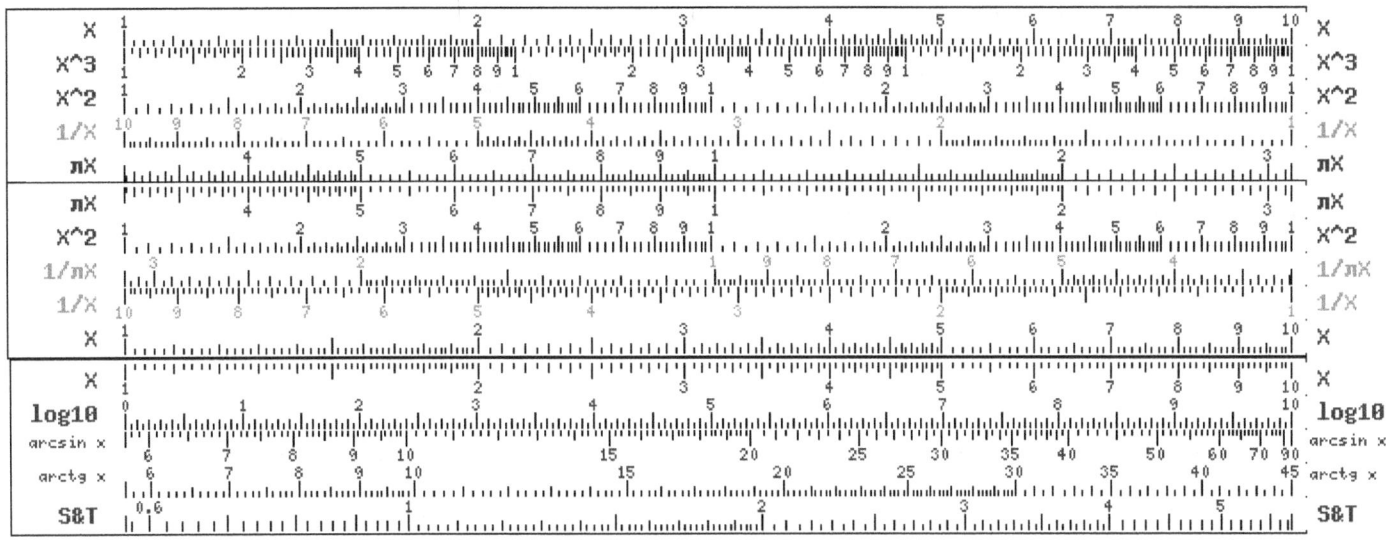

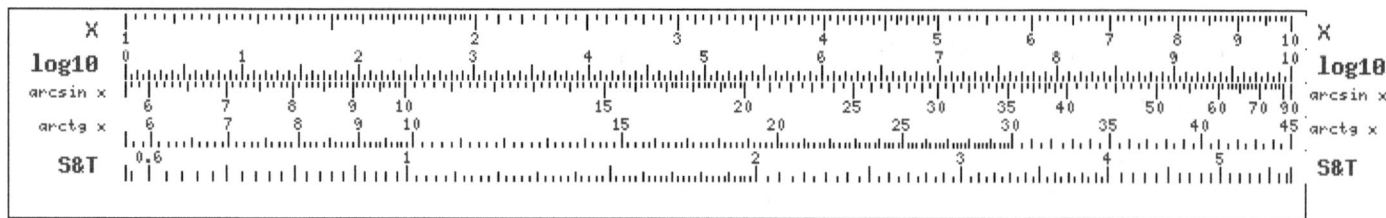

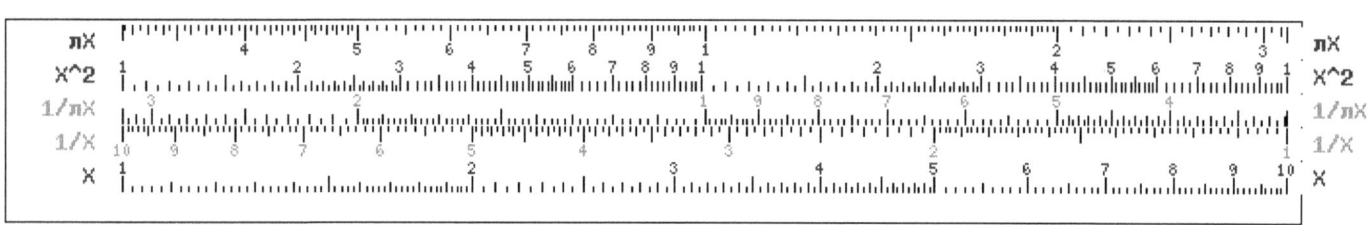

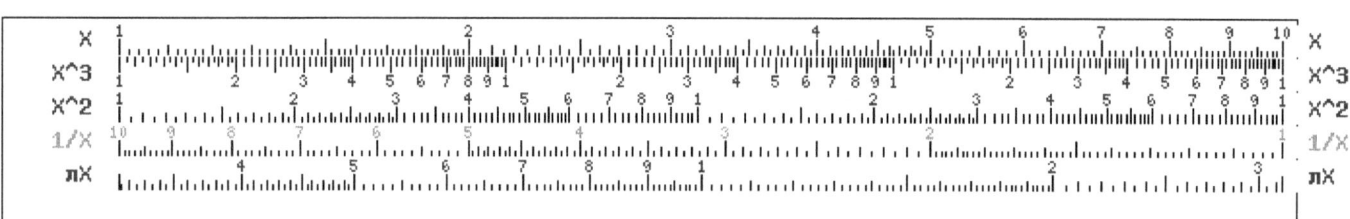